Advanced Manufacturing Methods

Advanced Manufacturing Methods: Smart Processes and Modeling for Optimization describes developments in advanced manufacturing processes and applications considering typical and advanced materials. It helps readers implement manufacturing 4.0 production techniques and highlights why a consolidated source and robust platform are necessary for implementing machine learning processes in the manufacturing sector.

- Discusses the industrial impact of manufacturing process
- Provides novel fundamental manufacturing solutions
- Presents the various aspects of applications in advanced materials in correlation of physical properties with macro-, micro- and nanostructures
- Reviews both classical and artificial manufacturing when applied with typical and novel innovative materials

Aimed at those working in manufacturing, mechanical and optimization of manufacturing processes, this work provides readers with a comprehensive view of current development in, and applications of, advanced manufacturing.

Advanced Manufacturing Methods

Smart Processes and Modeling for Optimization

Edited by
Catalin I. Pruncu and Jamal Zbitou

CRC Press
Taylor & Francis Group
Boca Raton London New York

CRC Press is an imprint of the
Taylor & Francis Group, an **informa** business

First edition published 2023
by CRC Press
6000 Broken Sound Parkway NW, Suite 300, Boca Raton, FL 33487-2742

and by CRC Press
2 Park Square, Milton Park, Abingdon, Oxon, OX14 4RN

© 2023 Taylor & Francis Group, LLC

CRC Press is an imprint of Taylor & Francis Group, LLC

ISBN: 978-0-367-37089-3 (hbk)
ISBN: 978-1-032-32636-8 (pbk)
ISBN: 978-0-367-82238-5 (ebk)

DOI: 10.1201/9780367822385

Typeset in Times
by KnowledgeWorks Global Ltd.

Contents

Preface vii
Editor Biographies xi
List of Contributors xiii

Chapter 1   Enabling Smart Manufacturing with Artificial Intelligence
                and Big Data: A Survey and Perspective 1

                *Huu Du Nguyen, Kim Phuc Tran, Philippe Castagliola,
                and Fadel M. Megahed*

Chapter 2   Green Applications with an Advanced Manufacturing
                Method: Cold Spray Deposition Technology 27

                *Koray Kılıçay, Salih Can Dayı, Esad Kaya,
                and Selim Gürgen*

Chapter 3   Multi-Criteria Decision-Making Applications in Conventional
                and Unconventional Machining Techniques 57

                Şenol Bayraktar and Erhan Şentürk

Chapter 4   Taguchi-Based GRA Method for Multi-Response
                Optimization of Spool Bore in EHSV Made Up
                of Stainless Steel 440C 83

                Pranav R., Md. I. Equbal, Azhar Equbal, and Kishore K.

Chapter 5   Wearing Behaviour of Electrodes during EDM
                of AISI 1035 Steel 101

                Azhar Equbal and Md. Asif Equbal

Chapter 6   Achieving Optimal Efficiency in Manufacturing
                through Reinforced PA 3D Printed Parts
                Generated by FDM Technology 113

                *Aissa Ouballouch, Rachid El Alaiji, Mohammed Sallaou,
                Aboubakr Bouayad, Hamza Essoussi, Said Ettaqi,
                and Larbi Lasri*

Chapter 7   Characterization of Effect of Cellular Support Structures
               in Selective Laser Melting Using Stainless Steel 316L.................... 143

               M. Abattouy, M. Ouardouz, and H. Azzouzi

Chapter 8   Using Carbon Nanotubes for Advanced Manufacturing
               of Antibiofilm Materials... 159

               M. Gomes, R. Teixeira-Santos, L. C. Gomes, and F. J. Mergulhão

Chapter 9   Development of a Novel Nanocomposite Coating
               for Tribological Applications 175

               Arti Yadav, Muthukumar M, and M. S. Bobji

Index.. 195

Preface

Today, the application of artificial intelligence (AI) is very broad, and its increase was noted especially in areas like understanding of natural language, visual recognition, robotics, autonomous system, machine learning, design and manufacturing and other critical fields. AI has evolved massively, thanks in particular to the emergence of Cloud Computing and Big Data that are capable of storing and linking the enormous amount of data from the critical sector for improving daily life such as manufacturing. The present book will highlight the recent progress in fundamental research in advanced manufacturing methods, integrating various aspects from synthesis to applications of advanced materials and providing a correlation of physical properties with macro, micro and nanostructures, which is a great interest for the academic and industrial readers. Moreover, it will provide a cutting-edge research from around the globe in this field. Current status, trends, future directions, opportunities, etc., will be discussed, making it friendly for beginners and young researchers. This book will present and discuss new studies that incorporate some modern techniques such as AI, multi-criteria decision and novel advanced material from macro to microscale representative of a society which main desire is to achieve net zero emission by 2050. The state of art of each research field was presented briefly and concisely in each chapter. It makes this book a desired tool for the university in order to accommodate the new students with novel knowledge in advanced manufacturing that are based in AI and novel material. It can also be useful for training courses, engineers, PhD students and other researchers.

This book is composed of nine chapters:

Chapter 1 deals with manufacturing incorporating AI and Big Data, especially for the development of Industry 4.0. This chapter is organized as follows: firstly, we find an introduction about the use of internet of things (IoT) technologies in specific industrial applications such as factories, manufacturing, facilities and warehouses. The IoT (like RFID technology) can create new business models by improving productivity, exploiting analytics for innovation, maximizing operational efficiency, optimizing business operations and protecting systems. Secondly, were presented a part reserved for Big Data from sensors and IoT devices. As known, a large number of special sensors are used to collect data in smart manufacturing, in which devices are independent of each other. The fourth part of this chapter is about the application of AI in the industry, which makes the manufacturing sector more smart. The AI algorithms are able to learn from data; enhance themselves by learning new heuristics. After the presentation of the concepts, applications and current researches of the IoT, Big Data, and AI for smart manufacturing in the Industry 4.0 era are revealed. We have indicated opportunities and further research perspectives of these important factors of industrial intelligent manufacturing. In the last section, we presented a case study linked to the application of AI algorithms for the remaining useful lifetime (RUL) prognostic of a product.

Chapter 2 discusses the use of cold spray (CS) technology in detail and its green applications are examined. CS technology is a thermal spray coating method in which dense coating layers can be produced with a high deposition rate. Deposition layers are produced by the mechanical locking mechanism, which is formed as a result of the impact of the powder particles sprayed from the nozzle at supersonic speeds on the substrate. It is defined as cold since the accelerated sprayed powder particles are solid state due to the process temperature below powder melting temperature. Thanks to the important advantages it provides, this method can also be used in advanced manufacturing methods such as additive manufacturing and innovative methods such as repairing damaged parts. For this reason, it is defined in advanced manufacturing methods and can also be evaluated in the green applications class due to its environmental effects.

Chapter 3 is focused on the application of multi-criteria decision making (MCDM) techniques when is applied to conventional and non-conventional machining techniques. The use of MCDM methods is based on modeling and analysis of decision processes according to specific criteria. MCDM is preferred as an assist tool in determining the most appropriate performance conditions by reducing cost and time in manufacturing. This chapter discusses the different techniques VIKOR, COPRAS, MULTIMOORA, WASPAS, EVAMIX, OCRA and MABAC, which are still being developed and current studies on processing in the literature related to these techniques are presented in detail and comparatively. The VIKOR method is widely preferred in the literature in studies on machinability, COPRAS covers qualitative and quantitative features and allows the selection of alternatives among the results. MULTIMOORA is used for a ratio system in which the response of the alternative on a target is compared with a denominator representing all the alternatives of the target. WASPAS is a multi-response appropriate decision-making method. EVAMIX, which is among the decision-making analysis approach systems, reduces the selection time or decision-making process. OCRA is used to calculate the performance of alternatives in performance and efficiency measurement and analysis problems. And for the MABAC method, it is used to determine the most suitable alternatives.

Chapter 4 discusses and presents the optimal WEDM process for machining of the spool bore, which is a critical component in EHSV. The dimension of spool bore must be precisely controlled in order to maintain the smooth functioning of the servo system. Therefore, there is required to have an optimized process input to the WEDM process for precise and controlled machining of spool bore. The study contains a detailed experimental investigation that was performed to obtain an optimal combination of process parameters for the WEDM of a spool bore for a type II EHSV. The spool bore is made by machining of stainless steel of grade 440C. Experiments were designed in accordance with Taguchi array. The significance of parameters affecting the quality characteristics is established using ANOVA. Grey-based Taguchi method is employed for multi-response optimization method.

Chapter 5 is concentrated on the study of electrical discharge machining (EDM), which is a precise machining technique where machining is done by a series of repetitive sparks between electrode and workpiece. Since the performance of electrodes is important in EDM, it is necessary to understand the complex wearing behavior of electrodes obtained during machining. In this study, it was concluded that the wear of electrode is affected by a number of factors. By consequent, it is recommended to impose an appropriate control for the machining factors used in EDM. To understand that wear of copper electrode (tool wear ratio, TWR), the machining of AISI 1035 steel were used as case study. After the determination of significant EDM factors, affecting the TWR, an optimal set of factors that yield lower TWR was determined using the main effect plot and desirability function approach.

Chapter 6 deals with additive manufacturing (AM) technologies such as fused deposition modeling (FDM). In this chapter, we have provided a study on the effect of PPs when using the FDM technique on mechanical properties, dimensional accuracy, surface roughness and total cost is investigated. Hence, an experimental method for AM of chopped glass reinforced polyamide (GRPA) and chopped Kevlar reinforced polyamide (KRPA) is presented. These PPs include extrusion temperature (ET), layer thickness (LT) and print speed (PS). In the conducted study, a detailed investigation of performance and quality of 3D printed polyamide composites (with chopped glass fiber and Kevlar fiber reinforcement) was presented. The comparison of these composites with fabricated additively neat polyamide and ABS parts and those processed by injection molding was carried out. By consequent, the dimensional accuracy of both PAs was assessed and found to be influenced by extrusion temperature and layer thickness more than print speed and reinforcement. KRPA surface roughness was largely affected by process parameters more than GRPA. Total cost was found to be notably influenced by print speed, layer thickness and nature or reinforcement.

Chapter 7 discusses a full factorial design of experiment (DOE) for cone support, tree support and different cellular support structures manufactured from stainless steel 316L using selective laser melting for selected geometric control factors. Then digital microscopy is used, which enables to study of upper surface quality. The morphology of surface was further examined through cross sectioning and revealing the deformation mechanisms too. Afterward, a removability evaluation of every sample from the platform was investigated. The purpose of this study was to compare the features of two distinct types of support structures (tree and cellular supports). Each support type was manufactured from 316L stainless steel as control material.

Chapter 8 is focused on the field of surface engineering and advanced manufacturing, carbon nanotubes (CNTs) which have been drawing industrial attention not only due to their unique structural, mechanical and electrical properties but also to their antimicrobial activity. The high antimicrobial activity of CNT-nanocomposites was reported against a broad spectrum of microorganisms and their potential for medical and water treatment applications was demonstrated. Also, the significant fouling resistance of these nanocomposites was proven at distinct levels, including in the development of marine AF or FR coatings, water treatment, and industrial processes such as filtration.

Chapter 9 focuses on the manufacturing of nanocomposite coating by introducing the development of a novel nanocomposite coating with ordered porous alumina (NPA) as a matrix embedded with aligned metal (Cu) nanorods. This was achieved by optimally modifying the barrier layer without sacrificing the interfacial strength. Uniform coating has been achieved over a specific area. The coating is found to have good tribological properties of low friction and high wear resistance. Also, the use of pulse electrodeposition is a highly efficient and well-suited method for a metal filling into the pores.

Dr. Catalin Pruncu
Brunel University London
Uxbridge UB8 3PH, UK

Dr. Jamal Zbitou
Laboratory (LABTIC) – ENSA of Tangier
University of Abdelmalek Essaâdi
Morocco

Editor Biographies

Dr. Catalin Pruncu is an Associate Lecturer at Brunel University London, UK and Visiting Researcher at Politecnico di Bari, Italy. He is a former Researcher Fellow in the Design, Manufacturing and Engineering Management at the University of Strathclyde, Glasgow, UK. Catalin has more than ten years of research experience in academia and industry. Dr. Pruncu published more than 100 papers in ISI journals, 3 books, a patent and other papers at various national and international conferences. Dr. Pruncu is a Charter and Member of the Institute of Mechanical Engineers (UK) since November 2015. He has experience in prestigious universities (Imperial College London, University of Birmingham, University of Sussex) and industries such as IMI Truflo Marine Ltd. and Spanish Navy. Recently he was invited as Editor for Special Issue, Wear Behavior of Polymer Composites and Mathematical Modeling and Simulation in Mechanics and Dynamic Systems, MDPI and also he is a reviewer for almost 50 ISI journals including *Measurement*, *Elsevier*, *Journal of Materials Research and Technology*, *Surface and Coatings Technology*, *Journal of Cleaner Production*, etc. He was involved in organizing different international conferences including the 12th International Conference on New Trends in Fatigue and Fracture, Brasov, Romania, 2012.

Dr. Jamal Zbitou was born in Fes, Morocco, in June 1976. He received his PhD degree in Electronics from Polytech of Nantes, the University of Nantes, France, in 2005. He is currently a Professor of Electronics in ENSA of Tangier, University of Abdelmalek Essaâdi, Morocco and an Associate Professor at the University of Quebec in Outaouais, Canada. He is involved in the design of hybrid, monolithic active and passive microwave electronic circuits and also involved in the design of Rectennas, RFiD Tag and their applications in wireless communications and wireless power transmission (WPT).

List of Contributors

M. Abattouy
University of Abdelmalek Essaâdi
Tétouan, Morocco

Rachid El Alaiji
University of Abdelmalek Essaâdi
Tétouan, Morocco

H. Azzouzi
University of Abdelmalek Essaâdi
Tétouan, Morocco

Şenol Bayraktar
Recep Tayyip Erdoğan University
Rize, Turkey

M. S. Bobji
Indian Institute of Science
Bangalore, India

Aboubakr Bouayad
University of Moulay Ismail
Meknes, Morocco

Philippe Castagliola
Université de Nantes
Nantes, France

Salih Can Dayı
Eskişehir Osmangazi University
Eskişehir, Turkey

Azhar Equbal
Faculty of Engineering and Technology
Jamia Millia Islamia
New Delhi, India

Md. Asif Equbal
Cambridge Institute of Technology
Ranchi, India

Md. I. Equbal
University Polytechnic, Aligarh Muslim
    University
Aligarh, U.P., India

Hamza Essoussi
University of Moulay Ismail
Meknes, Morocco

Said Ettaqi
University of Moulay Ismail
Meknes, Morocco

L. C. Gomes
University of Porto
Porto, Portugal

M. Gomes
University of Porto
Porto, Portugal

Selim Gürgen
Eskişehir Osmangazi University
Eskişehir, Turkey

Kishore K.
Vasavi College of Engineering
Ibrahim Bagh, Hyderabad, India

Esad Kaya
Eskişehir Osmangazi University
Eskişehir, Turkey

Koray Kılıçay
Eskişehir Osmangazi University
Eskişehir, Turkey

Larbi Lasri
University of Moulay Ismail
Meknes, Morocco

Muthukumar M
Indian Institute of Science
Bangalore, India

Fadel M. Megahed
Farmer School of Business
Miami University
Miami, OH, USA

F. J. Mergulhão
University of Porto
Porto, Portugal

Huu Du Nguyen
Institute of Artificial Intelligence
and Data Science
Dong A University
Danang, Vietnam

M. Ouardouz
University of Abdelmalek Essaâdi
Tétouan, Morocco

Aissa Ouballouch
University of Hassan II
Casablanca, Morocco

Pranav R.
Maturi Venkata Subba Rao Engineering
College
Hyderabad, India

Mohammed Sallaou
University of Moulay Ismail
Meknes, Morocco

Erhan Şentürk
Recep Tayyip Erdoğan University
Rize, Turkey

R. Teixeira-Santos
University of Porto
Porto, Portugal

Kim Phuc Tran
GEMTEX - Génie et Matériaux Textiles
Univ. Lille, ENSAIT
Lille, France

Arti Yadav
Indian Institute of Science
Bangalore, India

1 Enabling Smart Manufacturing with Artificial Intelligence and Big Data

A Survey and Perspective

Huu Du Nguyen[1], Kim Phuc Tran[2], Philippe Castagliola[3], and Fadel M. Megahed[4]

[1]Institute of Artificial Intelligence and Data Science, Dong A University, Danang, Vietnam
[2]Univ. Lille, ENSAIT, ULR 2461 - GEMTEX - Génie et Matériaux Textiles, F-59000 Lille, France
[3]Nantes Université & LS2N UMR CNRS 6004, Nantes, France
[4]Farmer School of Business, Miami University, Miami, Ohio, United States

CONTENTS

1.1 Introduction 2
1.2 The Industrial Internet of Things 3
1.3 Big Data from Sensors and IIoT Devices 5
1.4 Artificial Intelligence in Industrial Applications....................................... 6
1.5 Perspectives for IIoT, Big Data, and AI in the Smart Manufacturing 9
    1.5.1 Monitoring Production Process 9
    1.5.2 Product Design 9
    1.5.3 Product Lifecycle Management 10
    1.5.4 Predictive Maintenance 11
    1.5.5 Cybersecurity.. 11
    1.5.6 Manufacturing Optimization.. 12
    1.5.7 Virtual Reality in a Smart Manufacturing 12
    1.5.8 Machine-to-Machine Communication 13
    1.5.9 Wearable Technology and Smart Manufacturing............................. 13

DOI: 10.1201/9780367822385-1

1

1.6   A Case Study 14
      1.6.1 C-MAPSS Datasets 14
      1.6.2 The Proposed AI-Based Method for RUL with the C-MAPSS
             Datasets.. 15
      1.6.3 Experimental Results.. 15
1.7   Conclusion 19
References..20

1.1   INTRODUCTION

A path through the four industrial revolutions represents stages in the development of industrial systems from manual work toward smart manufacturing. The use of water and steam-powered mechanical manufacturing facilities is considered as the first industrial revolution. The next revolution was the discovery of electricity and assembly line production based on the division of labor. The third industrial revolution is the introduction of electronics and IT to production systems to enhance the automation of manufacturing. Recently, engineers have realized that manufacturing has been developed into a new era where products tend to control their manufacturing processes, leading to the concept of "Industry 4.0". The term Industry 4.0 has become increasingly pervasive in the context of industrial manufacturing, and it has been considered as the fourth industrial revolution (Henning[1]).

In Industry 4.0, by integrating advanced technologies like industrial internet of things (IIoT), Big Data, Cloud Computing, and artificial intelligent (AI), the manufacturing will become intelligent and independently perform complex tasks such as predicting, maintaining the machines, monitoring, and controlling the production. It is now at the center of Industry 4.0, and it attracts a lot of interest from governments, enterprises, and researchers. The framework for implementing smart manufacturing of Industry 4.0 was proposed in Frank et al.[2] An extensive review of technologies for smart manufacturing systems is recently conducted in Alcácer and Cruz-Machado.[3]

The IIoT refers to the use of internet of things (IoT) technologies to enhance manufacturing and industrial processes. Using the IIoT, the industrial manufacturing process and industrial products (components, machines) are connected to the Internet; the underlying equipment resources are integrated, leading to the abilities of perception, interconnection, and data integration of the manufacturing systems. The application of the IIoT in smart manufacturing could lead to a decrease in production costs by 10–30%, logistic costs by 10–30%, and quality management costs by 10–30% (Rojko[4]).

A vital characteristic of the IIoT is that the sensors are embedded in all the components related to the manufacturing process. These sensors act as the "eyes" for collecting data from the supply, production, storage, distribution, and consumption of products. With the ever-accelerating advancement of IoT devices and other communication and sensing devices and technologies, it is expected that the data generated from future smart manufacturing systems will grow exponentially, leading to the concept of Big Data (Qin[5]). The Big Data in smart manufacturing mainly encompasses real-time sensor data and manufacturing process data which have a large volume, multiple sources, and spare value. The applications of Big Data are rapidly developing in industrial supply chain analysis and optimization, product quality control, and active maintenance (Song et al.[6]; Xu et al.[7]).

The term "Big Data" does not simply refer to a huge amount of data but also the various kinds of collected data in all the stages. The massiveness, complexity, and heterogeneity of data streams require the advanced computing technologies, which are now performed efficiently thanks to the availability of AI. This is a set of algorithms related to the creation of machine intelligence that is able to perform tasks heretofore only performed by people (Fox[8]). It enables automatic processing of data toward highly complex feature abstraction instead of handcrafting the optimum feature representation of data with domain knowledge. In the past, the computer was programmed to perform a specific task. Now, the AI makes the computer intelligent with the ability to correctly interpret external data, learn from such data, and use those learnings to achieve specific goals and tasks through flexible adaptation (Kaplan and Haenlein[9]). The use of AI can revolutionize the industrial manufacturing process with a large number of applications such as predictive maintenance, predictive quality analytics, automation, and insightful identification of engineering systems.

It brings countless advantages to Industry 4.0, involving optimizing all stages of the manufacturing process, reducing waste, and creating new smart products and services with high quality. The AI now plays the role of a "brain" for smart manufacturing.

Toward an intelligent manufacturing industry is a long-term and not straightforward process. It requires a deep insight into a multiplicity of advanced and modern technologies that are integrated into this process. This study aims to provide a survey of the key techniques that enable smart manufacturing, including IIoT, Big Data, and AI. Several important perspectives for these techniques in smart manufacturing will be discussed and suggested. Some obtained results in this chapter have already been discussed in Nguyen et al.[10]

The chapter is organized as follows: The IIoT-based background and techniques are presented in Section 1.2. In Section 1.3, we focus on the concept of Big Data in Industry 4.0 and the methods of Big Data analytics. The industrial applications of the AI algorithms are discussed in Section 1.4. Section 1.5 is devoted to several important perspectives for further research on the application of IIoT, Big Data, and AI in the smart manufacturing sector. A case study is given in Section 1.6. Section 1.7 provides some concluding remarks.

1.2   THE INDUSTRIAL INTERNET OF THINGS

The IIoT can be simply understood as the use of IoT technologies in specific industrial applications such as factories, manufacturing facilities, and warehouses. In practice, it is broadly similar to and interchangeable with Industry 4.0. More general and comprehensive definitions of the IIoT have been discussed in Boyes et al.[11] Based on a large number of modern technologies such as cyber-physical systems (CPS), Cloud Computing, mobile technologies, machine-to-machine (M2M), advanced robotics, IoT, radio frequency identification (RFID) technology, and cognitive computing, the IIoT incorporates machine learning algorithms and Big Data technologies, harnessing the sensor data, M2M communication, and automation technologies that have existed in industrial settings for years (Saturno et al.[12]). As a result, the IIoT can create new business models by improving productivity, exploiting analytics for innovation, maximizing operational efficiency, optimizing business operations, and protecting systems. For example, according to a specific application of IIoT to the fashion industry provided in Shen et al.,[13] an innovative RFID-embedded smart washing machine has been proposed to produce a large amount of real-time data associated with what color, textile, style, and brands of clothes the consumers are washing, and when and where they are washed. This real-time information, which reflects what consumers are wearing, enables fashion companies to derive optimal solutions in terms of consumer preference and use analytics to improve operations performance in design, manufacturing, and retailing. The advantages of IIoT in intelligence manufacturing and smart factory are discussed broadly in the literature, see, for example, Zhong et al.[14] and Lu and Weng.[15]

The first essential basic platform for the IIoT is the IoT, which is a network of physical devices embedded with sensors, actuators, electronics, software, and network connectivity that enable these objects to connect, interact, and exchange data. The IoT is the bridge between the digital domain, involving a novel analytical approach, and the physical domain. Applications of the IoT can be seen in many areas such as consumer (smart home, elder care), commercial (medical and health-care, transportation, building, and home automation), and infrastructure (metro-politan scale deployments, energy management, environmental monitoring), and especially industry. Studies on IoT in the literature are abundant. Key technologies, applications, visions, and challenges of the IoT have been presented in Miorandi et al.[16] and Gubbi et al.[17] Five different categories of the solutions of IoT, including smart wearable, smart home, smart city, smart environment, and smart enterprise, have been discussed in a survey on the IoT marketplace from an industrial perspective conducted in Perera et al.[18] Yang et al.[19] provided a comprehensive review of the IoT for smart manufacturing. The authors outlined in their study the evolution of the Internet from computer networks to human networks to the latest era of smart and connected networks of manufacturing things. Recently, the arising of novel communication infrastructures is also the contribution to the IIoT. As the data generated by IIoT systems continues to increase exponentially, industrial companies face the challenge of transferring their critical communication infrastructure to enable digital and automated operations. New generation of networked, information-based technologies like 5th generation (5G) wireless mobile communication and low-power wide-area (LPWA) networks are expected to provide the means to allow an all-connected world of humans and objects. Surveys of long-range wireless technologies for IoT applications can be seen in Sinha et al.[20] Cheng et al.[21] analyzed the crucial technologies and difficulties of the 5G-based IIoT.

Another state-of-the-art technology of the IIoT is the CPS, a new class of auto-mated systems that enables the connection of the operations of physical objects with computing and communication infrastructures. The IoT can be referred to as the con-nection of the CPS to the Internet (Jazdi[22]). A CPS consists of a control unit, which controls the sensors, actuators that are necessary to interact with the real processes, and a large number of sensors. It contains networked interactions that are designed and developed with physical input and output, along with their cyber-twined services such as control algorithms and computational capacities (Zhong et al.[14]). The 5C architecture, i.e. smart connection, data-to-information conversion, cyber, cognition, and configuration, has been proposed in Lee et al.[23] This 5C structure provides a step-by-step guideline to construct a CPS system from the data acquisition to value creation. The CPS is now considered as one of the most significant advances in the development of computer science, information, and communication technologies which have a great impact on the emergence of new technologies like robotic surgery, intelligent buildings, smart electric grid, and implanted medical devices (Monostori et al.[24]). A typical example of CPS is the smart vehicle, where various types of sensors enable to collect raw data during the vehicle's operation, involving a driver's operation, vehicle condition, driving route, and destination. Herterich et al.[25] investigated the influence of CPS on industrial services in manufacturing. Waschull et al.[26] analyzed the impact of the technological capabilities provided by CPS on work design and they developed a detailed framework of the transformation toward CPS. A comprehensive survey of the CPS, involving various examples, several defining characteristics, design techniques, and its applications, has been introduced in Khaitan and McCalley.[27] The development of advanced IoT and CPS technologies brings to the IIoT not only enormous advantages, but it also reveals the challenges related to security vulnerabilities. The network connection of the CPS caries processes and exchanges a huge number of security-critical and privacy-sensitive data. The IIoT-based manufacturing systems are now one of the top industries targeted by a variety of attacks. Many security incidents affecting industrial control systems and critical infrastructure have been reported in Levy.[28] The problem of protecting IIoT systems against cyberattacks is becoming increasingly important and indispensable in their design. A mathematical framework for attack detection and identification in CPS and a brief review of the studies related to the analysis of vulnerabilities of CPS to external attacks are presented in Pasqualetti et al.[29] Sadeghi et al.[30] presented an overview of security and privacy challenges as well as possible solutions for IIoT systems, including designing security architectures for CPS, verifying the integrity of CPS, and securing IoT device management.

1.3   BIG DATA FROM SENSORS AND IIoT DEVICES

As mentioned above, a large number of special sensors are used to collect data in smart manufacturing, in which devices are independent of each other. The sensors turn the physical conditions of an object into an electrical signal. These electrical signals are then transferred to a programmable logic controller for further operations. Other electronic devices like an RFID chip, which is an electromagnetic technology for transferring data to detect and track tags of objects in automatic identification, are also widely used (Khan et al.[31]). Moreover, each component has the capability of communicating and sharing data based on new network technologies, for instant, wireless sensor networks. The rapid development of modern technologies using IIoT makes the process of data acquisition and storage increasingly easy and convenient. As a result, data at different stages of a product's life, ranging from raw materials, machines' operations, and facility logistics, are collected.

These various sources of sensors, machine log files, event streams from IoT devices, human activities, and industrial robotics promote the era of industrial Big Data. In the literature, the first efforts to define Big Data focused on enlisting its characteristics, leading to "3V", namely, Volume, Velocity, and Variety (McAfee et al.[32]). Volume refers to the size of data being collected from all the sources. Velocity informs the frequency of data acquisition; it is related to real-time data, data stream, operation data, remote, and control. The higher is the velocity, the higher is the volume. Variety describes the different types of data that may be handled, involving structured, semi-structured, and unstructured data. A variety of data directly affects their integrity: the more variety is in the data, the more errors it will contain. These characteristics have been extended by adding multiple features like Veracity (related to the unreliability or uncertainty of some data sources), Validity (the correctness and accuracy of data concerning the intended usage), Volatility (related to the retention policy of structured data), and Value (the desired outcome of Big Data processing) (Uddin et al.[33]). Other dimensions of Big Data like Vision, Validation, and Variability have also been recently mentioned in Alcácer and Cruz-Machado.[3] A general definition of Big Data based on a survey of existing definitions for this concept is given in De Mauro et al.[34]

The core feature of Big Data is that it requires to be analyzed, i.e. Big Data analytics, and without being analyzed, Big Data has no value. Big Data analytics refers to the process of collecting data, transferring data into centralized cloud data centers, preprocessing data, analyzing data, and visualizing data. Involving complex applications with elements such as information technology, mathematics, statistics, machine learning algorithms, and data mining techniques, Big Data analytics is the major contributor to enrich the intelligence of businesses. The use of Big Data analytics results in a 15–20% increase in return on investment for retailers (Perrey et al.[35]). In the literature, methods for data-driven anomaly detection for industrial Big Data have been studied in Martí et al.[36] and Zhang et al.[37] Xu et al.[7] presented an advanced fault diagnostic method to handle collected industrial Big Data. Two applications of Big Data analytics in manufacturing were mentioned in Chen et al.,[38] including active maintenance and product design optimization. Megahed and Jones-Farmer[39] provided several statistical perspectives on Big Data, in which the authors discussed many Big Data applications to highlight the opportunities and challenges for applied statisticians interested in surveillance and statistical process control. Challenges and opportunities for the implementation of Big Data analytics in the Industry 4.0 have been discussed in Khan et al.[31]

1.4   ARTIFICIAL INTELLIGENCE IN INDUSTRIAL APPLICATIONS

The key factor that makes smart manufacturing "smart" is artificial intelligence (AI). In the field of computer science, AI refers to the intelligence demonstrated by machines. It is a set of algorithms that enables a machine to perform complex tasks by perceiving the working environment and taking actions to maximize the possibility of successfully achieving the predetermined goals. The AI algorithms are able to learn from data; enhance themselves by learning new heuristics (Domingos[40]), and provide powerful tools to extract useful information and the connection or feature

TABLE 1.1

The Application of AI Algorithms in Industrial Manufacturing

Application Situation	Reference
Decision making support systems	42–43
Fault diagnosis	50–55
Predictive analytics	56–58
Advanced Robotics	59–65
Scheduling	67–72

from data that could not be analyzed effectively heretofore. Nowadays, the term "artificial intelligence" is ubiquitous and its applications have been witnessed in a large number of areas of everyday life. In industrial manufacturing, AI also provides a significant improvement in designing an automatic robot, making decisions, monitoring and scheduling the production process, predictive maintenance, and analytics. It can be said that the manufacturing and factories of Industry 4.0 cannot be "intelligent" or "smart" without AI. In the literature, an extensive review of the application of AI to industrial manufacturing was carried out in Meireles et al.[41] Other advantages of AI in smart manufacturing have been discussed in Frank et al.[2] Table 1.1 provides typical application scenarios of the AI algorithms in smart manufacturing. In particular, Helu et al.[42] and Teti et al.[43] reviewed the machine learning techniques of neural networks and genetic algorithms for decision-making support systems. The advantages, challenges, and applications of typical techniques of machine learning like instance-based learning, Support Vector Machine, Random Forest, Bayesian Networks, and Artificial Neural Network have been illustrated in Wuest et al.[44] in a wide range of industrial applications. A survey of the advanced use and development of machine learning in smart manufacturing has recently been given in Sharp et al.[45]

In smart manufacturing, the mechanical systems are closely linked to form continuous production lines. Failure due to degradation or an abnormal working environment can lead to unexpected downtime and affect the productivity of the entire system. Two important applications of the AI algorithms can be used to deal with this problem, involving (1) diagnostic analytics for fault assessment and (2) predictive analytics for defect prognostic (Wang et al.[46]). For the first application, the fault diagnostics of several electronic systems and electric drives using machine learning have been presented in Fenton et al.[47] and Murphey et al.[48] The strengths and weaknesses of the machine learning method for fault prognostic are discussed in Gao et al.[49] A number of algorithms of deep learning have recently been investigated to overcome the limitations of traditional machine learning methods. A novel method for early fault detection of machine tools based on deep learning and dynamic identification has been proposed in Luo et al.[50] The Convolutional Neural Network (CNN), the Deep Belief Network (DBN), and the Stacked Auto Encoder have been applied to several fault diagnostics such as bearing, gearbox, aircraft engine, reciprocating compressor, rolling-element bearing, wind generator, and wind turbine; see, for example, Chen et al.[51]; Janssens et al.[52]; Lu et al.[53]; Li et al.[54]; and Sun et al.[55] The second application makes maintenance intelligent as it allows to predict the appropriate time for performing maintenance. The remaining useful life estimation of machinery using DBN and long short-term memory (LSTM) network are given in Deutsch et al.[56] The deep learning algorithms such as the DBN and the Support Vector Regression have been proposed for predictive analytics in Malhi et al.[57] and Wang et al.[58] A comprehensive review of the application of deep learning algorithms for smart manufacturing has been conducted in Wang et al.[46]

Another algorithm of the AI which has surfaced as a method with a great impact on Industry 4.0 is reinforcement learning (RL). It is an area of machine learning concerned with the problem of a software agent that tries to develop a behavioral strategy in order to maximize some notion of cumulative reward as a result of taking the right actions in any state of its environment. An RL agent interacts with its environment in discrete time steps, and basic reinforcement is modeled as a Markov decision process. The major contribution of RL to smart manufacturing is perhaps in robotics, a core feature of Industry 4.0. Kober et al.[59] conducted a survey about the use of the RL algorithm in operating industrial robots for studies before 2013. Studies in Levine et al.[60] and Mnih et al.[61] used deep RL to tackle a wide variety of motion planning for industrial robots directly from sensory input. These studies are then improved to avoid the computational effort of the learning problem in Meyes et al.[62] Deep RL with a smooth policy update is also applied to robotic cloth manipulation in Tsurumine et al.[63] Pane et al.[64] introduced two RL-based compensation methods for robot manipulators and then evaluated the proposed algorithms on a 6-DoF industrial robotic manipulator arm to follow different kinds of reference paths or to track a trajectory on a three-dimensional (3D) surface. The problem of advanced planning for autonomous vehicles in traffic has been solved by using RL and deep inverse RL in You et al.[65] Moreover, the application of RL in smart manufacturing is not only limited to robotics. The performance, stability, and deep approximators of RL for control have been reviewed in Busoniu et al.[66] The authors explained how to approximate representations of the solution make RL feasible for problems with continuous states and control actions. The RL algorithms also contribute significantly to scheduling, which is the process of arranging, controlling, and optimizing work and workloads in a production process or manufacturing process. A large number of research on the problem of scheduling have been conducted in the literature, for example, Xie et al.[67]; Fu et al.[68]; and Leusin et al.[69]

In smart manufacturing, since the CPS is integrated with the production, logistics, and services, it is crucial to consider real-time scheduling. According to a discussion in Shiue et al.,[70] there are two main approaches in real-time scheduling from previous studies, involving the multi-pass simulation and machine learning approaches (e.g. artificial neural networks, decision tree learning, and support vector machine). The authors have pointed out several drawbacks of these two methods and then proposed an RL-based model for real-time scheduling for a smart factory. The idea of using deep RL for global production scheduling is also presented in Waschneck et al.[71] Other applications of AI in real scheduling problems can be seen in Costantino et al.[72]

1.5 PERSPECTIVES FOR IIoT, BIG DATA, AND AI IN THE SMART MANUFACTURING

The previous sections present the concepts, applications, and current research of the IIoT, Big Data, and AI for smart manufacturing in the Industry 4.0 era. In this section, we discuss some opportunities and further research perspectives on these important factors in industrial intelligent manufacturing.

1.5.1 MONITORING PRODUCTION PROCESS

The monitoring production process is an important problem in smart manufacturing. Recently, machine learning algorithms have been proposed within Statistical Process Monitoring (SPM), which has been shown to effectively detect a variety of abnormal conditions. This approach converts the monitoring problem to an outlier detection problem or a supervised classification problem, which classifies future observations as either in- or out-of-control. When a huge amount of data are collected, it makes sense to use SPM techniques to analyze them in order to get accurate information about special causes of variation. The idea of using One-Class Support Vector Machines for detecting abnormality in Tran et al.[73] can be developed for further application. The technology significantly reduces the time taken for testing and designing new prototypes and the duration for redesigning the existing models. In addition, with the rapid development of IIoT technologies, data are measured with a high frequency, high dimension, and a large variety which should not be treated straightforwardly. Therefore, it is necessary to develop new methods to be adapted to monitor these Big Data. The Topological Data Analysis (TDA) very recently emerged as a powerful tool to extract insights from high-dimensional, incomplete, and noisy data of varying types such as images, 3D scans, graphs, point clouds, and meshes. The core idea of TDA is to find the shape, the underlying structure of shapes, or relevant low-dimensional features of high-dimensional data. As a result, the problem of treating the complex structure and massive data is brought to a simpler problem. Among some recent research on TDA, the first successful application of TDA in the manufacturing systems domain is presented in Guo and Banerjee.[74] In this study, the authors apply the Mapper algorithm, one of the tools of the TDA field, for predictive analysis of a chemical manufacturing process data set for yield prediction and a semiconductor etch process data set for fault detection. In general, there is still little research on this promising approach in the literature and further research needs to be conducted to discover the possible numerous applications to smart manufacturing. For example, deep learning algorithms such as LSTM and CNN for topological data should be developed to monitor smart manufacturing processes.

1.5.2 PRODUCT DESIGN

One of the indispensable requirements for enterprises is to create new products to meet the increasingly specialized demands of customers. Making the product development process, which drives new products from the idea of production to the final product that is launched on the market, with less cost and fewer mistakes, is the key to the competitiveness of a manufacturing company. The new products need to have the ability to update based on customers' preferences. For designers, this means thinking ahead to feature consumers in order to add them to the new products. This process requires to continuously update and predict the user's preference, usage context, product features, as well as their interrelations. The traditional approaches to collect information about users' experiences such as questionnaires, surveys, and self-reports with predefined questions and prompts are recently replaced by the social media platform-based methods and other advanced technologies of IIoT. On the one hand, it helps customers easier to make choices more easily. On the other hand, it collects user-generated reviews. This big customer-generated data require AI algorithms to be analyzed to deeply understand the users' experience and extract them to assist product design. The AI algorithms ensure the active integration of customers' knowledge into new product development, improving quality as well as reducing product development time and costs. A survey on AI and expert systems application in new product development was conducted early in Rao et al.[75] Shen et al.[13] presented a specific example of color trend forecasting using an extreme learning machine forecasting model. This forecast of color trends is an important piece of advice for designers to design new fashion models: the new prototype will more likely meet the customers' demand. A comprehensive overview of the methods to exploit the user experience from the online customer for product design is conducted in Yang et al.[76] The authors also design a semi-supervised learning approach to classify the candidate segments. Other applications of machine learning methods in innovative product development can be seen in Zhan et al.[77] In addition to meeting the market demand, product recovery is also an important factor in the design of the product, especially for end-of-life products. The IIoT is a very useful technology to perform this task. The sensors and RFID embedded tags allow these products to be recovered via disassembly to meet the components' demands, remanufactured to meet the product demands, or recycled to meet the demands of the materials. The study of evaluating different designs of a product for the ease of disassembly and remanufacturing based on IIoT presented in Joshi and Gupta[78] can be developed for further applications.

1.5.3 PRODUCT LIFECYCLE MANAGEMENT

Product lifecycle management (PLM) refers to the succession of strategies for managing all data related to the design, production, support, and ultimate disposal of manufactured goods. From this point of view, monitoring the production process and product design could be considered as the components of PLM. The PLM brings tremendous benefits to manufacturing industries for improving product quality, reducing prototyping costs, identifying potential sales opportunities and revenue contributions, maintaining operational serviceability, and reducing environmental impacts at end-of-life. A study in Venkatasubramanian[79] reported that the petrochemical industry in the U.S. incurs approximately $20 billion in losses due to poor management of equipment and processes, which lead to such abnormal situations. Similarly, U.S. manufacturers spend over $7 billion annually recalling and renewing over 2000 defective products. All of these costs are associated with PLM. A key factor in PLM is prognostic and diagnostic monitoring. By comparing various existing approaches for prognostic and diagnostic in PLM, Venkat Venkatasubramanian[79] concluded that no traditional single method is adequate to handle all the requirements for a desirable diagnostic system. The application of the framework of the intelligent system for this complex problem is then suggested. In this sense, the use of IIoT, Big Data, and AI is a basic platform to design such intelligent PLM systems. The IIoT-based PLM system will be the first place where all product information from marketing and design comes together while the AI algorithms process this information and output them in a form suitable for production and support. As an example, recently, Karasev and Sukhanov[80] designed a PLM system using multi-agent systems models. In general, developing AI-based PLM systems in smart manufacturing is attractive for future research.

1.5.4 PREDICTIVE MAINTENANCE

Another application of IIoT, Big Data, and AI in intelligent manufacturing is predictive maintenance or just-in-time maintenance. The benefits and advantages to be achieved by the development of comprehensive predictive maintenance are shown in Ferreiro et al.[81] Currently, the maintenance is regularly scheduled at fixed intervals, leading to some limitations. It could be a large amount of money in lost productivity while the process must be halted to fix the failure if a machine breaks before the maintenance. Otherwise, time and another amount of money would be wasted if the machine does not need any maintenance at the moment. Moreover, unnecessary maintenance operations increase the failure rate because of installed defective items or human negligence. Advanced IIoT technologies allow engineering to carry out predictive maintenance. In particular, the sensors are applied to a different piece of equipment to continuously update the individual equipment health. The AI tools can process this gathered data to monitor and forecast overloads, machinery failures, or related problems based on learning algorithms. The appropriate time that the machines need to be maintained is determined exactly. Online learning, transfer learning, and domain adaption are the trends for predictive maintenance and prognosis in Industry 4.0 (Diez-Olivan et al.[82]).

1.5.5 CYBERSECURITY

The exponential growth of IIoT also brings a significant challenge in designing and implementing smart factories related to the cybersecurity problem. It would result in severe and heavy consequences if hackers could gain access to control network or malware and worms could invade and destroy the operating system of a factory. Cybersecurity is now a major concern in many types of research. Twenty-four risk assessment methods developed for or applied in the context of a Supervisory Control and Data Acquisition (SCADA) system have been selected and examined in Cherdantseva et al.[83] Tuptuk and Hailes[84] provided a large number of real reported attacks against smart manufacturing systems as well as existing active and passive countermeasures with their limitations. In general, many traditional existing cybersecurity solutions have become obsolete (Mahmood and Afzal[85]). Machine learning and deep learning algorithms for regression, classification, and clustering are powerful tools to detect and identify different classes of network attacks. For example, the method for network intrusion detection based on nested one-class support vector machines presented in Nguyen et al.,[86] the method for anomaly detection in time series based on LSTM networks presented in Bontemps et al.[87] could be applied and developed for further applying to cybersecurity in intelligent manufacturing.

1.5.6 MANUFACTURING OPTIMIZATION

Optimizing the production process is a key mission in smart manufacturing. It could be an action, a process, or a methodology to make operations related to the manufacturing process as good, functional, and effective as possible. This is a highly complex task where the best combination of a large number of controllable parameters must be found to achieve a predetermined goal. The IIoT and AI platform enable the smart factory to gather, store, and process data related to products from raw materials, production, and use process to customer feedback. The features extracted and the insight from these data and the AI optimization algorithms are extremely necessary to reach optimality in the manufacturing process. By refining this process and employing programs that can upgrade existing systems, track and report errors, and allow for experiments with design, the AI enables the smart factory to establish robust systems, which can maximize the yield in a shorter time and at reduced costs. The RL algorithm is applied to optimize the inventory in Oroojlooyjadid et al.[88] Combined with simulation techniques such as multi-agent modeling, the AI can provide a decision support system for supplier selection and production planning based on real-time data (Waschneck et al.[71]). The AI algorithms also can predict long-term production demands and transform them into daily production orders, considering last-minute orders and operations' restrictions (Frank et al.[2]). In general, the question of how to optimize each stage as well as the whole of the manufacturing process is still a major concern in designing smart manufacturing.

1.5.7 VIRTUAL REALITY IN A SMART MANUFACTURING

Virtual Reality (VR) can be considered an artificial environment created by a mixture of interactive hardware and software to present users with realistic 3D images as a real environment in which she/he may interact within a seemingly real or physical way. In the past, the application of VR in industrial manufacturing could only be seen in Robotics, where VR technology has been used to control robots in telepresence and telerobotic systems. Recently, the rapid development of AI algorithms has transformed the use of VR in intelligent manufacturing. VR technology allows us to design and test virtual prototypes without the necessity to build physical prototypes. It is applied to support factory layout planning in several industries such as aerospace, trucks, snus, and tobacco (Gong et al.[89]) and maintenance in aeronautics (Ceruti et al.[90]). Advances in VR technology present a new opportunity that can provide the implementation of complex engineering theory from industrial real-life practice in a virtual 3D model (Ma et al.[91]). The reasons why VR can be a powerful tool for applications in smart manufacturing have been discussed in Hamid et al.,[92] where the authors pointed out three specific applications of VR in industrial manufacturing, including Design (design and prototyping), Manufacturing Processes (machining, assembly, and inspection), and Operation Management (planning, simulation, and training). A review of the studies related to virtual manufacturing systems can be seen in Dobrescu et al.[93] Although the AI algorithms (machine learning and deep learning) have now made significant advances in VR for hand and eye-tracking gestures, natural language processing, detailed environmental mapping, VR can better simulate an environment by replicating one that is already existing. They can use external structure sensors with an AI system to create a mixed reality experience for their users and incorporate voice commands into training simulations. It can be said that there are still many promising applications of VR technology that can be developed for the application of smart manufacturing.

1.5.8 MACHINE-TO-MACHINE COMMUNICATION

As a vital part of the use of IIoT in smart manufacturing, a M2M communication technology represents technologies that allow two or more devices to exchange information and data with each other. That is, the communication between machines is autonomous without human intervention. In fact, the applications of this technology span across various areas such as healthcare, remote monitoring, security, and city automation. In manufacturing, digital control systems and smart sensors can maximize operational efficiency, safety, and reliability. A comprehensive survey of opportunities for M2M communication is conducted by Amodu and Othman.[94] The M2M technology is also the basic technical background for a large number of applications in Industry 4.0 like robotics and automation. Obviously, the recent development of IIoT technologies and AI algorithms can significantly enhance the use of M2M in smart manufacturing.

1.5.9 WEARABLE TECHNOLOGY AND SMART MANUFACTURING

A promising technology emerged in recent times that benefiting smart manufacturing is wearable technology. It describes electronics and computers that are integrated into clothing and other accessories that can be worn comfortably on the body. In practice, one may have seen the applications of this technology in everyday life such as fitness trackers, continuous health monitoring, smart clothes, and contactless payment solutions. In the literature, a large number of studies have been devoted to the development of smart wearable devices based on the IoT, Big Data, and AI techniques (Maman et al.[95,96]; Nguyen et al.[97]). In the field of manufacturing, the use of smart wearable devices can create great benefits. Firstly, it helps to improve productivity and efficiency in manufacturing. For instance, the wearable devices can be integrated with voice functions to become hands-free instructions and communication devices. Some specific industries like oil, gas, and automotive industries often have complicated instructions. With smart wearable devices, workers can stay focused on their tasks, obtain additional information, or deliver remote commands (Cavuoto et al.[98]). Secondly, it enhances safety in the manufacturing process, especially in high-tech intelligent manufacturing. This technology can create a safer working environment for workers by monitoring ambient conditions and alarming potential accidents; or it can allow workers to perform more accurate operations in the manufacturing process, leading to better products. In addition, it also increases the authentication and security planning. As maintaining proper security protocols and authenticate employees is mandatory in some factories or processes, wearable devices can take the form of authentication devices that restrict access to a facility. Due to promising applications of wearable technology in assisting smart manufacturing, some studies related to the topic have been introduced in the literature. Roda-Sanchez et al.[99] introduced OperaBLE, an IoT-based Wearable to improve efficiency and smart worker care services in Industry 4.0. The role of wearable devices in meeting the needs of cloud manufacturing was discussed in Hao and Helo.[100] Among smart wearable devices, smart glasses and smartwatches are the two best-known ones. A state of the art of wearable devices and the corresponding industrial applications can be seen in Aleksy et al.[101] In general, the rapid advances in the IIoT technique and AI algorithms can benefit wearable technology, enabling further applications in smart manufacturing.

1.6   A CASE STUDY

In this section, we present an application of AI algorithms to the remaining useful lifetime (RUL) prognostic. In prognostics and health management, the RUL represents the amount of time left before equipment is considered to not perform its operation. An accurate prognostic of the RUL plays a very important role in any manufacturing process: it enables the manufacturers to assess accurately an equipment's health status. As a result, they will have better plan logistics and minimize the costs involved by timely conducting maintenance activities. Due to its importance, a large number of studies has been devoted to this problem. Especially, the recent advancements of IIoT techniques and AI algorithms have benefited to the RUL prognostic as the obtained results have become increasingly accurate. Following this direction, we suggest a combination of the Kalman filter, the CNN, and the LSTM algorithms to create a novel model for the RUL based on the C-MAPSS datasets.

1.6.1   C-MAPSS Datasets

The Commercial Modular Aero-Propulsion System Simulation (C-MAPSS) datasets were generated with the C-MAPSS simulator, a tool for the simulation of realistic large commercial turbofan engine data. These datasets are composed of four distinct datasets (FD001, FD002, FD003, and FD004), containing information from different aircraft gas turbine engines. Each dataset is further divided into training and test sets of multiple multivariate time series. The training set is with run-to-failure information and the testing is with information terminating before a failure is observed. The data provided are from a high-fidelity system-level engine simulation that is designed to simulate fault engine degradation over several flights. The C-MAPSS datasets are well-known and they are widely used for several studies with different purposes in the literature. More details of the datasets can be seen in Costa and Kaymak[102]; the main information of the datasets is reproduced in Table 1.2.

TABLE 1.2
The C-MAPSS Datasets

Dataset	FD001	FD002	FD003	FD004
No. of training engines	100	260	100	249
No. of testing engines	100	259	100	248
Operating conditions	1	6	1	6
Fault modes	1	1	2	2

1.6.2 THE PROPOSED AI-BASED METHOD FOR RUL WITH THE C-MAPSS DATASETS

The structure of our proposed model is shown in Figure 1.1. At the first stage, a Kalman filter is used as a noise filtering to account for uncertainty from raw unlabeled input data of the C-MAPSS datasets since they are contaminated with sensor noise and lack of specific information on the effects of operational conditions. Then, an LSTM layer with 100 neurons in total is added in the second stage to reveal hidden information and learn long-term dependencies in sequential data with multiple operating and fault conditions. Following this LSTM is the CNN layer, which is used to extract local features. It contains a 1D convolution layer, a 1D Maxpooling layer and it employs rectified linear units (ReLU) to calculate the feature maps. Another LSTM layer with 50 neurons is included in the third stage for the extraction of long-range dependencies features. This LSTM layer is to enhance the efficiency of the proposed method. The ability of the LSTM to learn more patterns in data over long sequences makes them suitable for multivariate time series forecasting. In the final stage, there is a time distributed fully connected output layer to handle error calculations and perform RUL predictions. To keep the transparency of the chapter, we do not present the detail of each algorithm here, involving the Kalman filter, the CNN algorithm, and the LSTM algorithm. In the literature, references for these algorithms are abundant. Readers can refer to Harvey[103] and Goodfellow et al.[104] for a deeper understanding of these algorithms.

1.6.3 EXPERIMENTAL RESULTS

In the C-MAPSS datasets, engines are supposed to start in good condition and they begin to degrade at some points during the time series. In the training sets, the degradation increases in magnitude until failure, while in the test sets, it sometimes ends before failures. That means in the test sets, the last time step for each engine provides information on the true RUL targets. Thus, the main objective is

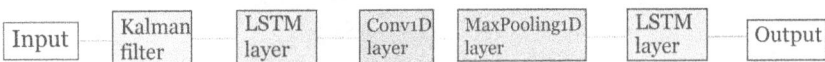

FIGURE 1.1 A graphical architecture of the proposed model.

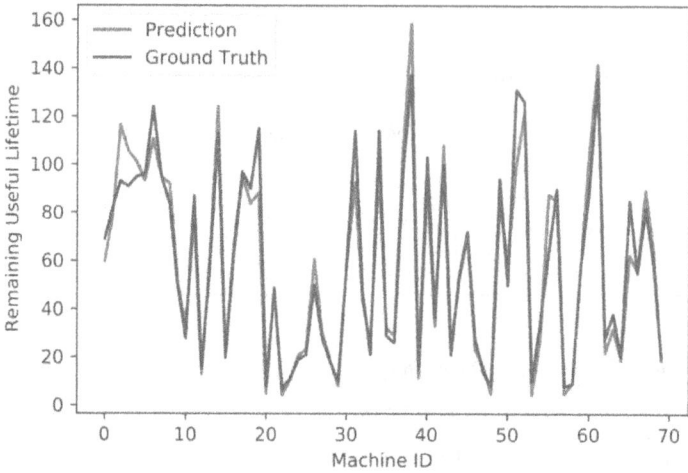

FIGURE 1.2 The true RUL and the predicted RUL by the proposed approach on the FD001 dataset.

now to predict the correct RUL value for each engine in the test sets. We apply our proposed method for the RUL prediction based on the first three datasets in the C-MAPSS datasets, i.e. FD001, FD002, and FD003. The fourth dataset, i.e. FD004, is passed to avoid a too lengthy chapter. The predicted RUL on the first datasets and the corresponding line plot of train loss and validation loss from the proposed model during training are shown in Figures 1.2 and 1.3. From Figure 1.3, we can see that the model has comparable performance on both train and validation

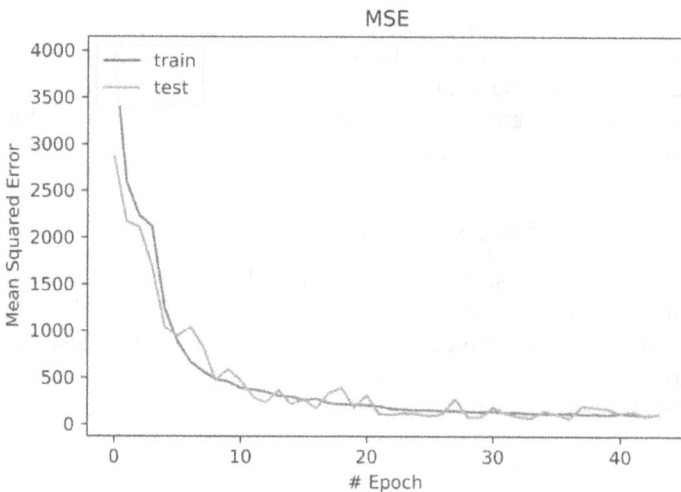

FIGURE 1.3 Line plot of train and validation loss from the proposed model during training on the FD001 dataset.

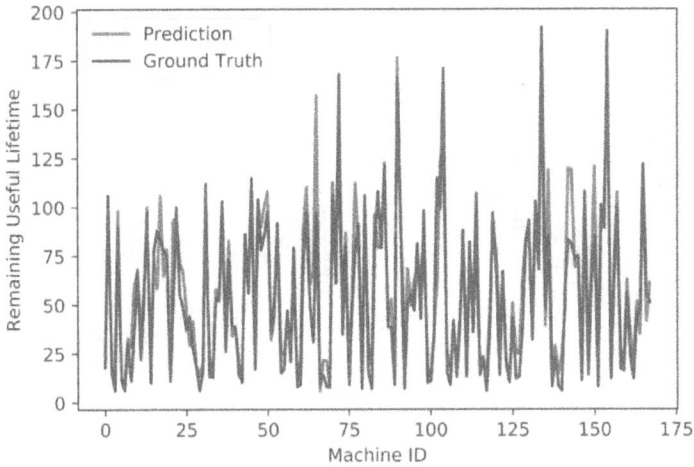

FIGURE 1.4 The true RUL and the predicted RUL by the proposed approach on the FD003 dataset.

datasets (labeled test), as the two lines are very close after a few epochs. This is also the case for the third dataset, FD003, as shown in Figures 1.4 and 1.5. For FD002 datasets, the obtained results are presented in Figures 1.6 and 1.7, where its line plot of train and validation shows that the proposed method is less effective in this dataset compared to the first and the third ones. In addition, for all three datasets, the predicted RUL matches very well with the true RUL, indicating the effectiveness of RUL prediction of the proposed method.

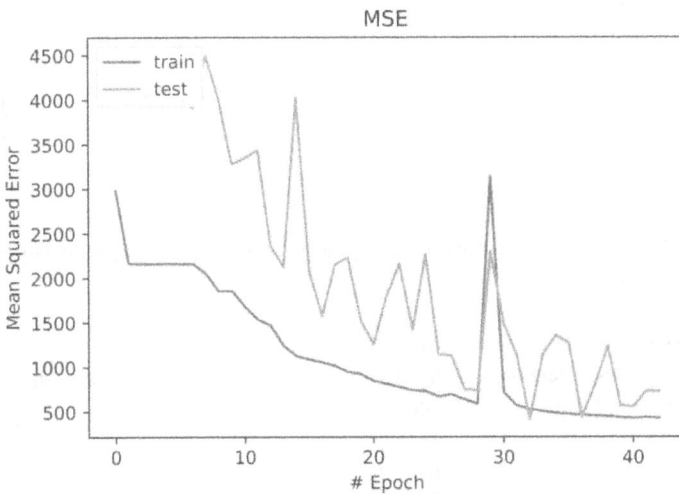

FIGURE 1.5 Line plot of train and validation loss from the proposed model during training on the FD003 dataset

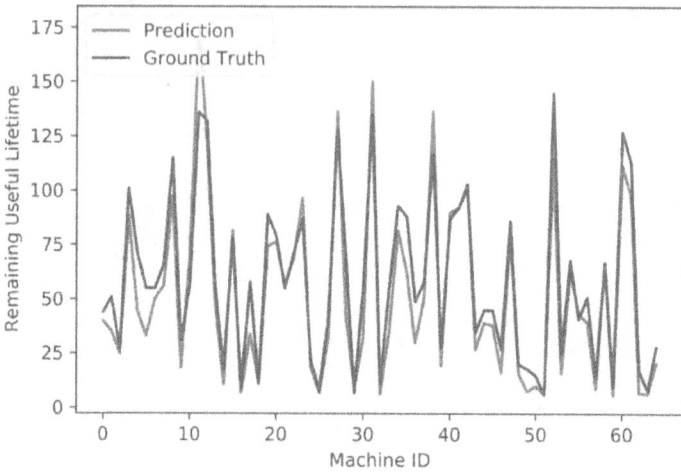

FIGURE 1.6   The true RUL and the predicted RUL by the proposed approach on the FD002 dataset.

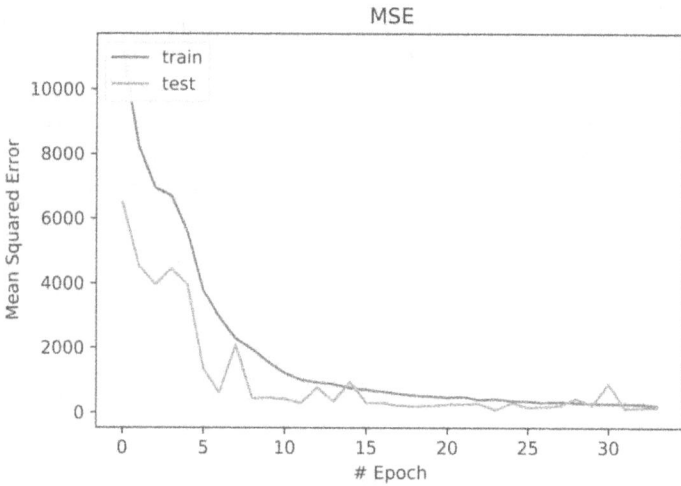

FIGURE 1.7   Line plot of train and validation loss from the proposed model during training on the FD002 dataset.

We also compared the performance of the proposed model and four very recent models available in the literature, using the root mean square error (RMSE). The RMSE is defined by the following formula

$$RMSE = \sqrt{\frac{1}{n}\sum_{i=1}^{n}d_i^2} \, , \qquad\qquad (1.1)$$

TABLE 1.3
RMSE Comparison with the Literature on the C-MAPSS Datasets

AI Approach & Refs.	FD001	FD002	FD003
Attention-based LSTM[107]	14.53		
MTW-BLSTM ensemble[105]	12.61		
LSTM- FW-CatBoost[108]	15.8	21.4	16.0
Semi-supervised RBM-LSTM-FNN[106]	12.56	22.73	**12.10**
Proposed method	**9.81**	**15.56**	13.05

where n is the total number of true RUL targets in the test set and $d_i = RUL_{predicted} - RUL_{true}$.

As discussed in Xia et al.,[105] the C-MAPSS FD001 dataset has been extensively used to verify the model in many studies. The authors also summarized the results of start-of-the-art RUL prediction models and stated that their proposed method achieved the best performance with the smallest RMSE equal to 12.61. This result is similar to the one obtained from the model in Ellefsen et al.[106] (with an RSEM = 12.56) and better than the one from Chen et al.[107] (with an RSEM = 14.53) and Deng et al.[108] (with an RSEM = 15.8). Compared to these models, our proposed method finds a significantly better result with an RSEM = 9.81. Our proposed method also leads to the smallest RSEM = 15.56 for the FD002 dataset. Meanwhile, the semi-supervised setup in Ellefsen et al.[106] achieved slightly higher RMSE prediction accuracy on the FD003 dataset. The compared results are presented in Table 1.3, where the best accuracies are in bold. From this result, we can say that our proposed model is a promising approach to achieve better accuracy in RUL prediction.

1.7 CONCLUSION

We have provided in this chapter a survey on enabling smart manufacturing with AI, IIoT, and Big Data. We have also discussed several perspectives and opportunities for the promising application of these key components in problems related to smart manufacturing. In general, the wide use of electronic sensing devices, wireless sensor networks, and other advanced technologies in the IIoT makes the process of collecting, transforming, and storing data from all stages of the manufacturing process easier and more convenient, promoting the era of big manufacturing data. Meanwhile, AI algorithms are used as powerful analytic approaches for insight into these data. The extracting insightful information and analyzing from big manufacturing data bring numerous benefits such as optimizing the production process, enhancing product quality, reducing cost, and making the manufacturing "smart". Finally, we have presented a case study in RUL prognostic where we have suggested an AI-based model achieve better accuracy of RUL prediction compared to other models existing in the literature. Despite a large number of researchers contributing to the application of these technologies to intelligent manufacturing, there are still many challenges for further research. No doubt that the new advanced inventions in IIoT, Big Data, and AI will make a decisive contribution to the development of Industry 4.0.

REFERENCES

1. Kagermann Henning. Recommendations for Implementing the Strategic Initiative INDUSTRIE 4.0: Securing the Future of German Manufacturing Industry; Final Report of the Industrie 4.0 Working Group, Forschungsunion, April 2013.
2. Alejandro Germán Frank, Lucas Santos Dalenogare, and Néstor Fabián Ayala. Industry 4.0 technologies: Implementation patterns in manufacturing companies. *International Journal of Production Economics*, 210:15–26, 2019.
3. Vitor Alcácer and Virgilio Cruz-Machado. Scanning the industry 4.0: A literature review on technologies for manufacturing systems. *Engineering Science and Technology, an International Journal*, 22(3):899–919, 2019.
4. Andreja Rojko. Industry 4.0 concept: Background and overview. *International Journal of Interactive Mobile Technologies (IJIM)*, 11(5):77–90, 2017.
5. S Joe Qin. Process data analytics in the era of big data. *AIChE Journal*, 60(9):3092–3100, 2014.
6. Zhiting Song, Yanming Sun, Jiafu Wan, and Peipei Liang. Data quality management for service-oriented manufacturing cyber-physical systems. *Computers & Electrical Engineering*, 64:34–44, 2017.
7. Yan Xu, Yanming Sun, Jiafu Wan, Xiaolong Liu, and Zhiting Song. Industrial big data for fault diagnosis: Taxonomy, review, and applications. *IEEE Access*, 5:17368–17380, 2017.
8. Mark S Fox. Industrial applications of artificial intelligence. *Robotics*, 2(4):301–311, 1986.
9. Andreas Kaplan and Michael Haenlein. Siri, Siri, in my hand: Who's the fairest in the land? On the interpretations, illustrations, and implications of artificial intelligence. *Business Horizons*, 62(1):15–25, 2019.
10. H Nguyen, K Tran, X Zeng, L Koehl, P Castagliola, and Pascal Bruniaux. Industrial internet of things, big data, and artificial intelligence in the smart factory: A survey and perspective. In *ISSAT International Conference on Data Science in Business, Finance and Industry*. Danang, Vietnam, 2019.
11. Hugh Boyes, Bil Hallaq, Joe Cunningham, and Tim Watson. The industrial internet of things (IIoT): An analysis framework. *Computers in Industry*, 101:1–12, 2018.
12. Maicon Saturno, Vinícius Moura Pertel, Fernando Deschamps, and E de FR Loures. Proposal of an automation solutions architecture for Industry 4.0. In *Proceedings of the 24th International Conference on Production Research*. ICPR Poznan, 2017.
13. Bin Shen, Xuemei Ding, Yanyan Wang, and Shuyun Ren. RFID-embedded smart washing machine systems in the big data era: Value creation in fashion supply chain. In *Fashion Supply Chain Management in Asia: Concepts, Models, and Cases*, pages 99–113. Springer, 2019.
14. Ray Y Zhong, Xun Xu, Eberhard Klotz, and Stephen T Newman. Intelligent manufacturing in the context of Industry 4.0: A review. *Engineering*, 3(5):616–630, 2017.
15. Hsi-Peng Lu and Chien-I Weng. Smart manufacturing technology, market maturity analysis and technology roadmap in the computer and electronic product manufacturing industry. *Technological Forecasting and Social Change*, 133:85–94, 2018.
16. Daniele Miorandi, Sabrina Sicari, Francesco De Pellegrini, and Imrich Chlamtac. Internet of things: Vision, applications and research challenges. *Ad Hoc Networks*, 10(7):1497–1516, 2012.

17. Jayavardhana Gubbi, Rajkumar Buyya, Slaven Marusic, and Marimuthu Palaniswami. Internet of things (IoT): A vision, architectural elements, and future directions. *Future Generation Computer Systems*, 29(7):1645–1660, 2013.

18. Charith Perera, Chi Harold Liu, and Srimal Jayawardena. The emerging internet of things marketplace from an industrial perspective: A survey. *IEEE Transactions on Emerging Topics in Computing*, 3(4):585–598, 2015.

19. Hui Yang, Soundar Kumara, Satish TS Bukkapatnam, and Fugee Tsung. The internet of things for smart manufacturing: A review. *IISE Transactions*, 51(11):1190–1216, 2018.

20. Rashmi Sharan Sinha, Yiqiao Wei, and Seung-HoonHwang. A survey on LPWA technology: LoRa and NB-IoT. *ICT Express*, 3(1):14–21, 2017.

21. Jiangfeng Cheng, Weihai Chen, Fei Tao, and Chun-Liang Lin. Industrial IoT in 5G environment towards smart manufacturing. *Journal of Industrial Information Integration*, 10:10–19, 2018.

22. Nasser Jazdi. Cyber physical systems in the context of industry 4.0. In *2014 IEEE International Conference on Automation, Quality and Testing, Robotics*, pages 1–4. IEEE, 2014.

23. Jay Lee, Behrad Bagheri, and Hung-An Kao. A cyber-physical systems architecture for Industry 4.0-based manufacturing systems. *Manufacturing Letters*, 3:18–23, 2015.

24. László Monostori, Botond Kádár, T Bauernhansl, S Kondoh, S Kumara, G Reinhart, O Sauer, G Schuh, W Sihn, and K Ueda. Cyberphysical systems in manufacturing. *CIRP Annals*, 65(2):621–641, 2016.

25. Matthias M Herterich, Falk Uebernickel, and Walter Brenner. The impact of cyber-physical systems on industrial services in manufacturing. *Procedia CIRP*, 30:323–328, 2015.

26. Sabine Waschull, Jos AC Bokhorst, Eric Molleman, and Johan C. Wortmann. Work design in future industrial production: Transforming towards cyberphysical systems. *Computers & Industrial Engineering*, 139:105679, 2019.

27. Siddhartha Kumar Khaitan and James D McCalley. Design techniques and applications of cyberphysical systems: A survey. *IEEE Systems Journal*, 9(2):350–365, 2015.

28. Elias Levy. Crossover: Online pests plaguing the offline world. *IEEE Security & Privacy*, (6):71–73, 2003.

29. Fabio Pasqualetti, Florian Dörfler, and Francesco Bullo. Attack detection and identification in cyber-physical systems. *IEEE Transactions on Automatic Control*, 58(11):2715–2729, 2013.

30. Ahmad-Reza Sadeghi, Christian Wachsmann, and Michael Waidner. Security and privacy challenges in industrial Internet of Things. In *Design Automation Conference (DAC), 2015 52nd ACM/EDAC/IEEE*, pages 1–6. IEEE, 2015.

31. Maqbool Khan, Xiaotong Wu, Xiaolong Xu, and Wanchun Dou. Big data challenges and opportunities in the hype of Industry 4.0. In *2017 IEEE International Conference on Communications (ICC)*, pages 1–6. IEEE, 2017.

32. Andrew McAfee, Erik Brynjolfsson, Thomas H Davenport, DJ Patil, and Dominic Barton. Big data: The management revolution. *Harvard Business Review*, 90(10):60–68, 2012.

33. Muhammad Fahim Uddin, Navarun Gupta, et al. Seven V's of big data understanding big data to extract value. In *2014 Zone 1 Conference of the American Society for Engineering Education (ASEE Zone 1)*, pages 1–5. IEEE, 2014.

34. Andrea De Mauro, Marco Greco, and Michele Grimaldi. What is big data? A consensual definition and a review of key research topics. In *AIP Conference Proceedings*, volume 1644, pages 97–104. AIP, 2015.

35. J Perrey, D Spillecke, and A Umblijs. Smart analytics: How marketing drives short-term and long-term growth. *McKinsey Quarterly*, 00425–3, 2013.

36. Luis Martí, Nayat Sanchez-Pi, José Manuel Molina, and Ana Cristina Bicharra Garcia. Anomaly detection based on sensor data in petroleum industry applications. *Sensors*, 15(2):2774–2797, 2015.
37. Liangwei Zhang, Jing Lin, and Ramin Karim. An angle-based subspace anomaly detection approach to high-dimensional data: With an application to industrial fault detection. *Reliability Engineering & System Safety*, 142:482–497, 2015.
38. Baotong Chen, Jiafu Wan, Lei Shu, Peng Li, Mithun Mukherjee, and Boxing Yin. Smart factory of Industry 4.0: Key technologies, application case, and challenges. *IEEE Access*, 6:6505–6519, 2018.
39. Fadel M Megahed and L Allison Jones-Farmer. Statistical perspectives on big data. In *Frontiers in Statistical Quality Control, 11*, pages 29–47. Springer, 2015.
40. Pedro Domingos. *The Master Algorithm: How the Quest for the Ultimate Learning Machine Will Remake Our World*. Basic Books, 2015.
41. Magali RG Meireles, Paulo EM Almeida, and Marcelo Godoy Simões. A comprehensive review for industrial applicability of artificial neural networks. *IEEE Transactions on Industrial Electronics*, 50(3):585–601, 2003.
42. Moneer Helu, Don Libes, Joshua Lubell, Kevin Lyons, and Katherine C Morris. Enabling smart manufacturing technologies for decision-making support. In *ASME 2016 International Design Engineering Technical Conferences and Computers and Information in Engineering Conference*, American Society of Mechanical Engineers, page 50084, 2016.
43. Roberto Teti, Krzysztof Jemielniak, Garret O'Donnell, and David Dornfeld. Advanced monitoring of machining operations. *CIRP Annals*, 59(2):717–739, 2010.
44. Thorsten Wuest, Daniel Weimer, Christopher Irgens, and Klaus-Dieter Thoben. Machine learning in manufacturing: Advantages, challenges, and applications. *Production & Manufacturing Research*, 4(1):23–45, 2016.
45. Michael Sharp, Ronay Ak, and Thomas Hedberg Jr. A survey of the advancing use and development of machine learning in smart manufacturing. *Journal of Manufacturing Systems*, 48:170–179,, 2018.
46. Jinjiang Wang, Yulin Ma, Laibin Zhang, Robert X Gao, and Dazhong Wu. Deep learning for smart manufacturing: Methods and applications. *Journal of Manufacturing Systems*, 48:144–156, 2018.
47. William G Fenton, T. Martin McGinnity, and Liam P Maguire. Fault diagnosis of electronic systems using intelligent techniques: A review. *IEEE Transactions on Systems, Man, and Cybernetics, Part C (Applications and Reviews)*, 31(3):269–281, 2001.
48. Yi Lu Murphey, M Abul Masrur, ZhiHang Chen, and Baifang Zhang. Model-based fault diagnosis in electric drives using machine learning. *IEEE/ASME Transactions On Mechatronics*, 11(3):290–303, 2006.
49. Robert Gao, Lihui Wang, Roberto Teti, David Dornfeld, Soundar Kumara, Masahiko Mori, and Moneer Helu. Cloud-enabled prognosis for manufacturing. *CIRP Annals*, 64(2):749–772, 2015.
50. Bo Luo, Haoting Wang, Hongqi Liu, Bin Li, and Fangyu Peng. Early fault detection of machine tools based on deep learning and dynamic identification. *IEEE Transactions on Industrial Electronics*, 66(1): 509–518, 2019.
51. ZhiQiang Chen, Chuan Li, and René-Vinicio Sanchez. Gearbox fault identification and classification with convolutional neural networks. *Shock and Vibration*, 2015, 2015.
52. Olivier Janssens, Viktor Slavkovikj, Bram Vervisch, Kurt Stockman, Mia Loccufier, Steven Verstockt, Rik Van de Walle, and Sofie Van Hoecke. Convolutional neural network based fault detection for rotating machinery. *Journal of Sound and Vibration*, 377:331–345, 2016.
53. Chen Lu, Zhenya Wang, and Bo Zhou. Intelligent fault diagnosis of rolling bearing using hierarchical convolutional network based health state classification. *Advanced Engineering Informatics*, 32:139–151, 2017.

54. Chuan Li, René-Vinicio Sanchez, Grover Zurita, Mariela Cerrada, Diego Cabrera, and Rafael E Vásquez. Gearbox fault diagnosis based on deep random forest fusion of acoustic and vibratory signals. *Mechanical Systems and Signal Processing*, 76:283–293, 2016.

55. Wenjun Sun, Siyu Shao, Rui Zhao, Ruqiang Yan, Xingwu Zhang, and Xuefeng Chen. A sparse auto-encoder-based deep neural network approach for induction motor faults classification. *Measurement*, 89: 171–178, 2016.

56. Jason Deutsch, Miao He, and David He. Remaining useful life prediction of hybrid ceramic bearings using an integrated deep learning and particle filter approach. *Applied Sciences*, 7(7):649, 2017.

57. Arnaz Malhi, Ruqiang Yan, and Robert X Gao. Prognosis of defect propagation based on recurrent neural networks. *IEEE Transactions on Instrumentation and Measurement*, 60(3):703–711, 2011.

58. Peng Wang, Robert X Gao, and Ruqiang Yan. A deep learning-based approach to material removal rate prediction in polishing. *CIRP Annals*, 66(1):429–432, 2017.

59. Jens Kober, J Andrew Bagnell, and Jan Peters. Reinforcement learning in robotics: A survey. *The International Journal of Robotics Research*, 32(11):1238–1274, 2013.

60. Sergey Levine, Chelsea Finn, Trevor Darrell, and Pieter Abbeel. End-to-end training of deep visuomotor policies. *The Journal of Machine Learning Research*, 17(1):1334–1373, 2016.

61. Volodymyr Mnih, Adria Puigdomenech Badia, Mehdi Mirza, Alex Graves, Timothy Lillicrap, Tim Harley, David Silver, and Koray Kavukcuoglu. Asynchronous methods for deep reinforcement learning. In *International Conference on Machine Learning*, PMLR, pages 1928–1937, 2016.

62. Richard Meyes, Hasan Tercan, Simon Roggendorf, Thomas Thiele, Christian Büscher, Markus Obdenbusch, Christian Brecher, Sabina Jeschke, and Tobias Meisen. Motion planning for industrial robots using reinforcement learning. *Procedia CIRP*, 63:107–112, 2017.

63. Yoshihisa Tsurumine, Yunduan Cui, Eiji Uchibe, and Takamitsu Matsubara. Deep reinforcement learning with smooth policy update: Application to robotic cloth manipulation. *Robotics and Autonomous Systems*, 112:72–83, 2019.

64. Yudha P Pane, Subramanya P Nageshrao, Jens Kober, and Robert Babuška. Reinforcement learning based compensation methods for robot manipulators. *Engineering Applications of Artificial Intelligence*, 78:236–247, 2019.

65. Changxi You, Jianbo Lu, Dimitar Filev, and Panagiotis Tsiotras. Advanced planning for autonomous vehicles using reinforcement learning and deep inverse reinforcement learning. *Robotics and Autonomous Systems*, 114:1–18, 2019.

66. Lucian Busoniu, Tim de Bruin, Domagoj Tolić, Jens Kober, and Ivana Palunko. Reinforcement learning for control: Performance, stability, and deep approximators. *Annual Reviews in Control*, 46:8–28, 2018.

67. Zhanpeng Xie, Chaoyong Zhang, Xinyu Shao, Wenwen Lin, and Haiping Zhu. An effective hybrid teaching–learning-based optimization algorithm for permutation flow shop scheduling problem. *Advances in Engineering Software*, 77:35–47, 2014.

68. Yaping Fu, Jinliang Ding, Hongfeng Wang, and Junwei Wang. Two-objective stochastic flow-shop scheduling with deteriorating and learning effect in Industry 4.0-based manufacturing system. *Applied Soft Computing*, 68:847–855, 2018.

69. Matheus Leusin, Enzo Frazzon, Mauricio Uriona Maldonado, Mirko Kück, and Michael Freitag. Solving the job-shop scheduling problem in the Industry 4.0 era. *Technologies*, 6(4):107, 2018.

70. Yeou-Ren Shiue, Ken-Chuan Lee, and Chao-Ton Su. Real-time scheduling for a smart factory using a reinforcement learning approach. *Computers & Industrial Engineering*, 125:604–614, 2018.

71. Bernd Waschneck, André Reichstaller, Lenz Belzner, Thomas Altenmüller, Thomas Bauernhansl, Alexander Knapp, and Andreas Kyek. Optimization of global production scheduling with deep reinforcement learning. *Procedia CIRP*, 72:1264–1269, 2018.

72. Francesco Costantino, Alberto Felice De Toni, Giulio Di Gravio, and Fabio Nonino. Scheduling mixed-model production on multiple assembly lines with shared resources using genetic algorithms: The case study of a motorbike company. *Advances in Decision Sciences*, 2014, 2014.

73. Kim Phuc Tran, Truong Thu Huong, et al. Data driven hyperparameter optimization of one-class support vector machines for anomaly detection in wireless sensor networks. In *2017 International Conference on Advanced Technologies for Communications (ATC)*, pages 6–10. IEEE, 2017.

74. Wei Guo and Ashis G Banerjee. Identification of key features using topological data analysis for accurate prediction of manufacturing system outputs. *Journal of Manufacturing Systems*, 43:225–234, 2017.

75. S Subba Rao, Abraham Nahm, Zhengzhong Shi, Xiaodong Deng, and Ahmad Syamil. Artificial intelligence and expert systems applications in new product development – a survey. *Journal of Intelligent Manufacturing*, 10(3–4):231–244, 1999.

76. Bai Yang, Ying Liu, Yan Liang, and Min Tang. Exploiting user experience from online customer reviews for product design. *International Journal of Information Management*, 46:173–186, 2019.

77. Yuanzhu Zhan, Kim Hua Tan, and Baofeng Huo. Bridging customer knowledge to innovative product development: A data mining approach. *International Journal of Production Research*, 57(20):1–16, 2019.

78. Aditi D Joshi and Surendra M Gupta. Evaluation of design alternatives of end-of-life products using internet of things. *International Journal of Production Economics*, 208:281–293, 2019.

79. Venkat Venkatasubramanian. Prognostic and diagnostic monitoring of complex systems for product lifecycle management: Challenges and opportunities. *Computers & Chemical Engineering*, 29(6):1253–1263, 2005.

80. VO Karasev and VA Sukhanov. Product lifecycle management using multi-agent systems models. *Procedia Computer Science*, 103:142–147, 2017.

81. Susana Ferreiro, Egoitz Konde, Santiago Fernández, and Agustín Prado. Industry 4.0: Predictive intelligent maintenance for production equipment. In *European Conference of the Prognostics and Health Management Society*, Bilbao, Spain, pages 1–8, 2016.

82. Alberto Diez-Olivan, Javier Del Ser, Diego Galar, and Basilio Sierra. Data fusion and machine learning for industrial prognosis: Trends and perspectives towards Industry 4.0. *Information Fusion*, 50:92–111, 2019.

83. Yulia Cherdantseva, Pete Burnap, Andrew Blyth, Peter Eden, Kevin Jones, Hugh Soulsby, and Kristan Stoddart. A review of cyber security risk assessment methods for SCADA systems. *Computers & Security*, 56:1–27, 2016.

84. Nilufer Tuptuk and Stephen Hailes. Security of smart manufacturing systems. *Journal of Manufacturing Systems*, 47:93–106, 2018.

85. Tariq Mahmood and Uzma Afzal. Security analytics: Big data analytics for cybersecurity: A review of trends, techniques and tools. In *2013 2nd National Conference on Information Assurance (NCIA)*, pages 129–134. IEEE, 2013.

86. Quoc Thong Nguyen, Truong Thu Huong, Kim Phuc, Minh Kha Nguyen, Philippe Castagliola, and Salim Lardjane. Nested one-class support vector machines for network intrusion detection. In *2018 IEEE Seventh International Conference on Communications and Electronics (ICCE)*, pages 7–12. IEEE, 2018.

87. Loïc Bontemps, James McDermott, et al. Collective anomaly detection based on long short-term memory recurrent neural networks. In *International Conference on Future Data and Security Engineering*, pages 141–152. Springer, 2016.

88. Afshin Oroojlooyjadid, MohammadReza Nazari, Lawrence Snyder, and Martin Takáč. A deep Q-network for the beer game: A reinforcement learning algorithm to solve inventory optimization problems. *arXiv preprint arXiv:1708.05924*, 2017.

89. Liang Gong, Jonatan Berglund, Asa Fast-Berglund, Björn Johansson, Zhiping Wang, and Tobias Börjesson. Development of virtual reality support to factory layout planning. *International Journal on Interactive Design and Manufacturing (IJIDeM)*, pages 1–11, 2019.

90. Alessandro Ceruti, Pier Marzocca, Alfredo Liverani, and Cees Bil. Maintenance in aeronautics in an Industry 4.0 context: The role of augmented reality and additive manufacturing. *Journal of Computational Design and Engineering*, 6(4):516–526, 2019.

91. Junfeng Ma, Raed Jaradat, Omar Ashour, Michael Hamilton, Parker Jones, and Vidanelage L Dayarathna. Efficacy investigation of virtual reality teaching module in manufacturing system design course. *Journal of Mechanical Design*, 141(1):012002, 2019.

92. Nur Suraya Sahol Hamid, Faieza Abdul Aziz, and Amir Azizi. Virtual reality applications in manufacturing system. In *2014 Science and Information Conference*, pages 1034–1037. IEEE, 2014.

93. Radu Dobrescu, Daniel Merezeanu, and Stefan Mocanu. Process simulation platform for virtual manufacturing systems evaluation. *Computers in Industry*, 104:131–140, 2019.

94. Oluwatosin Ahmed Amodu and Mohamed Othman. Machine-to-machine communication: An overview of opportunities. *Computer Networks*, 145:255–276, 2018.

95. Zahra Sedighi Maman, Mohammad Ali Alamdar Yazdi, Lora A Cavuoto, and Fadel M Megahed. A data-driven approach to modeling physical fatigue in the workplace using wearable sensors. *Applied Ergonomics*, 65:515–529, 2017.

96. Zahra Sedighi Maman, Ying-Ju Chen, Amir Baghdadi, Seamus Lombardo, Lora A Cavuoto, and Fadel M Megahed. A data analytic framework for physical fatigue management using wearable sensors. *Expert Systems with Applications*, 155: 113405, 2020.

97. Huu Du Nguyen, Kim Phuc Tran, Xianyi Zeng, Ludovic Koehl, and GuillaumeTartare. An improved ensemble machine learning algorithm for wearable sensor data based human activity recognition. In *Reliability and Statistical Computing*, pages 207–228. Springer, 2020.

98. Lora Cavuoto, Fadel Megahed, et al. Understanding fatigue and the implications for worker safety. In *ASSE Professional Development Conference and Exposition*. American Society of Safety Engineers, 2016.

99. Luis Roda-Sanchez, Celia Garrido-Hidalgo, Diego Hortelano, Teresa Olivares, and M Carmen Ruiz. Operable: an IoT-based wearable to improve efficiency and smart worker care services in Industry 4.0. *Journal of Sensors*, 6272793, 2018.

100. Yuqiuge Hao and Petri Helo. The role of wearable devices in meeting the needs of cloud manufacturing: A case study. *Robotics and Computer-Integrated Manufacturing*, 45:168–179, 2017.

101. Markus Aleksy, Mikko J Rissanen, Sylvia Maczey, and Marcel Dix. Wearable computing in industrial service applications. *Procedia Computer Science*, 5:394–400, 2011.

102. Alp Akcay Yingqian Zhang Costa, Paulo Roberto de Oliveira, and Uzay Kaymak. Remaining useful lifetime prediction via deep domain adaptation. *Reliability Engineering & System Safety*, 195(106682), 2020.

103. Andrew C Harvey. *Forecasting, Structural Time Series Models and the Kalman Filter*. Cambridge University Press, 1990.

104. Ian Goodfellow, Yoshua Bengio, and Aaron Courville. *Deep Learning*. MIT Press, 2016.

105. Tangbin Xia, Ya Song, Yu Zheng, Ershun Pan, and Lifeng Xi. An ensemble framework based on convolutional bi-directional LSTM with multiple time windows for remaining useful life estimation. *Computers in Industry*, 115:103182, 2020.

106. André Listou Ellefsen, Emil Bjørlykhaug, Vilmar Æsøy, Sergey Ushakov, and Houxiang Zhang. Remaining useful life predictions for turbofan engine degradation using semi-supervised deep architecture. *Reliability Engineering & System Safety*, 183:240–251, 2019.

107. Zhenghua Chen, Min Wu, Rui Zhao, Feri Guretno, Ruqiang Yan, and Xiaoli Li. Machine remaining useful life prediction via an attention based deep learning approach. *IEEE Transactions on Industrial Electronics*, 68(3):2521–2531, 2020.

108. Kunyuan Deng, Xiaoyong Zhang, Yijun Cheng, Zhiyong Zheng, Fu Jiang, Weirong Liu, and Jun Peng. A remaining useful life prediction method with long-short term feature processing for aircraft engines. *Applied Soft Computing*, 93:106344, 2020.

2 Green Applications with an Advanced Manufacturing Method
Cold Spray Deposition Technology

*Koray Kılıçay[1], Salih Can Dayı[1],
Esad Kaya[1], and Selim Gürgen[2]*
[1]Department of Mechanical Engineering, Eskişehir
Osmangazi University, Eskişehir, Turkey
[2]Department of Aeronautical Engineering, Eskişehir
Osmangazi University, Eskişehir, Turkey

CONTENTS

2.1   Introduction 28
2.2   Cold Spray Technology...29
    2.2.1   Low-Pressure Cold Spray Process...30
    2.2.2   High-Pressure Cold Spray Process 31
    2.2.3   Comparison of Cold Spray Technology
           and Thermal Spray Technology...31
2.3   The Effect of Main Process Parameters on Cold Spray 33
    2.3.1   Nozzle Design...33
           2.3.1.1   Nozzle Traverse Speed...33
           2.3.1.2   Nozzle Scanning Step 34
           2.3.1.3   Gas Temperature...34
           2.3.1.4   Standoff Distance and Spray Angle......................................35
           2.3.1.5   Nozzle Trajectory..36
    2.3.2   Powder Morphology 36
2.4   Green Applications by Using Cold Spray Technology 37
    2.4.1   Anti-Pathogenic Applications...38
    2.4.2   Repair Applications 40
    2.4.3   Additive Manufacturing Using Cold Spray Technology 45
2.5   Future Perspective 50
2.6   Conclusions...51
References...52

DOI: 10.1201/9780367822385-2

2.1   INTRODUCTION

Material deposition via the cold spray (CS) method is a relatively innovative method. It was developed at the Russian Academy of Sciences in the 1980s and patented in the United States and Europe in 1994 and 1995. However, developments progressed slowly until 2007. With the development and utilization of high-pressure equipment, CS technology also takes its place in the literature with names such as supersonic particle deposition and gas dynamic cold spraying. Not only is frequently used in coating applications but it is also used in the repair of damages in metallic and composite materials in various industrial areas including the production of additive manufacturing parts. Studies on the CS method have been increasing in recent years (Champagne and Helfritch 2014; Shushpanov 1994; Widener et al. 2018). The material properties that cannot be obtained by using traditional thermal spray methods can be easily achieved by the CS method. The damaged mechanical parts can be repaired without exposing the melting temperature of metallic powder. Thus, undesirable oxidation, cracks, residual stress, and inclusions are prevented (Gärtner et al. 2006). CS technology could be considered in the "Green Applications" grade thanks to the recovery of the part and time-saving in the industry for high part costs. In addition, CS technology is also used as an advanced manufacturing method such as additive manufacturing due to includes the advantages of additive manufacturing methods. For this reason, it can be evaluated as "Green Applications with Advanced Manufacturing Method" grade, where it provides repair, additive manufacturing, and production of anti-pathogenic materials.

CS systems are used either large stationary systems or small portable systems in many research institutions in including Europe, Asia, and America around the world. CS systems process by accelerating particles aerodynamically and hitting the substrate surface at high speed (Champagne and Helfritch 2016). The main components of the CS system are the supersonic nozzle, the powder feed unit, and the gas heating systems. High-pressure He, N_2, or air is used as powder carrier gas. At temperatures below the powder melting temperature, it is sprayed from a supersonic nozzle to the substrate at high speed (Ashokkumar et al. 2021).

Recently, the CS repair method has been making great progress thanks to regeneration and recovery applications. These developments provide enormous benefits in the industrial sector. This method, which supports environmental development, is evaluated in the "green techniques and methods" grade. Apart from material recycling, it comes to the forefront as an alternative way of renewing and making it reliable in the product life cycle. Considering the entire production process of the product, it is clear that this method saves on raw material and energy consumption. But more importantly, it eliminates the problems such as delay time, a stock shortage, and economic limitations that occur during the replacement of the damaged part. For these reasons, the CS repair method is economically satisfactory (Raoelison et al. 2017). The CS process provides a clean production environment without burning fossil fuels. Deposition on the substrate material is formed by the impact of low-temperature metal powder particles on the surface at high speed (Ashokkumar et al. 2021).

The main difference of CS technology from other methods is that it uses kinetic energy instead of thermal energy to deposit the sprayed particles. The sprayed particles are below the melting temperature, unlike other methods. For this reason, this method is called cold. CS technology has grown rapidly in recent years in the field of coating, repair, and additive manufacturing applications, especially in the aviation and automobile sectors due to its significant advantages (Yu et al. 2021).

This section describes green applications and advanced manufacturing methods that can be achieved by CS. Particularly focused on repair, additive manufacturing, and anti-pathogenic applications. Also, experimental, existing applications and innovative approaches that may lead to new applications are examined.

2.2 COLD SPRAY TECHNOLOGY

The CS process is a solid-state material deposition method. Micron-sized powders are sprayed onto a substrate at high speeds, resulting in an interaction between the particles and the surface. As a result of severe plastic deformation, the powder material is deposited on the surface. Acceleration of particles is achieved by the expansion of a pressurized and heated gas at the nozzle. The particles remain solid throughout the process. Due to this event, the process is called CS. Figure 2.1 shows the schematic representation of the material deposition technology by CS.

In this process, metallic powder particles are injected into a Laval-type nozzle, where they are accelerated to high velocities by a supersonic gas stream. The gas is heated, accelerates to sonic velocity in the throat region of the nozzle, and then the flow becomes supersonic as it expands in the separation section of the nozzle (Mach number ranges from 2 to 4). The gas used can be He, air, N_2, and a mixture of these. N_2 is one of the most preferred process gases because it does not cause oxidation and is relatively cheap. In the process, the reached maximum temperature of the gas is typically 700°C, with a pressure peak of 3.5 MPa. The material deposition rate varies between 300 and 1200 m/s. If the used gas is air, the speed can increase to 600 m/s. The powder sizes used vary between 1 and 50 μm. When the accelerated particles hit the layer, they transform their kinetic energy into mechanical energy which causes plastic deformation and accumulation on the surface (Davis 2004; Zhao et al. 2006).

The sprayed particles remain solid throughout the deposition process compared to conventional thermal sputtering techniques and expose high plastic deformation when they hit the surface. In this way, kinetic energy is used instead of thermal energy for deposition. Therefore, compared to conventional thermal spraying methods, oxidation is greatly reduced and phase transformation is not observed, and thermal stresses are greatly reduced (Ashokkumar et al. 2021).

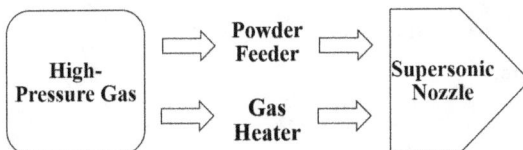

FIGURE 2.1   Schematic representation of CS technology.

Two different joint mechanisms occur in CS deposition processes. The first mechanism is the formation of the first layer by the powders sprayed on the surface. The formation of the first layer is related to the adhesion strength of the coating to the substrate surface. With the preliminary surface preparation processes, the surface can be made rough and the deposition efficiency can be increased. The second mechanism is the mechanical bonding of sprayed powders to each other. Due to the formation of more suitable layers at this stage, the thickness of the coating layer gradually increases. When sprayed powders hit the surface at high speed, they expose severe plastic deformation. This results in mechanical interlocking, causing the morphology of the coating powder to differ in size and shape of the particles embedded in the coating layer (Kılıçay 2020).

There are two different CS technology systems according to the operating gas pressure. These are low-pressure cold spray (LPCS, up to 9 bar) and high-pressure cold spray (HPCS, up to 40 Bar) systems. Although the main difference between these two systems is pressure, there are also differences in their functional principles in the inclusion of powders in the system.

2.2.1 Low-Pressure Cold Spray Process

As seen in Figure 2.2, in LPCS systems, powders enter the supersonic nozzle from a separate section. It is designed as a smaller and portable system as it does not require a high-pressure powder feeding system. These systems can be robot-controlled or manually controlled. Particle velocities are smaller than high-pressure systems. Typical particle velocities are limited to between 500 and 800 m/s. The sprayed powders must exceed the critical velocity value required for surface adhesion. For this reason, LPCS can be used in the application of mostly low-density materials. The gases used for particle accelerators are air, N_2, He, or mixtures thereof. Although helium gas provides the highest particle velocity, it is quite expensive compared to other gases. In cases where the sprayed particles are different sizes, large particles may not reach the critical velocity value. Therefore, larger and slower particles hit the substrate and bounce back from the surface. This situation deteriorates deposition efficiency. The deposition efficiency is defined as the ratio of the powder deposited on the substrate to the mass of the sprayed powder. The deposition efficiency in LPCS varies over a wide range from 10% to 95%. Powder splashed from the surface can be collected and reprocessed (Champagne and Helfritch 2016; Moridi et al. 2014).

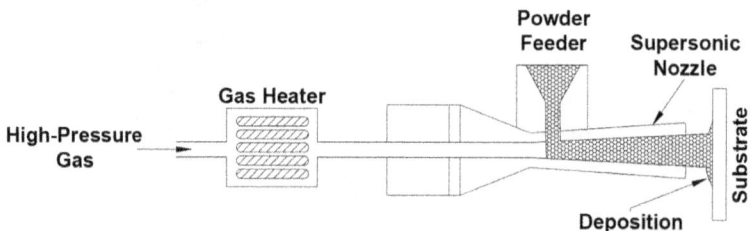

FIGURE 2.2   Schematic of the LPCS process.

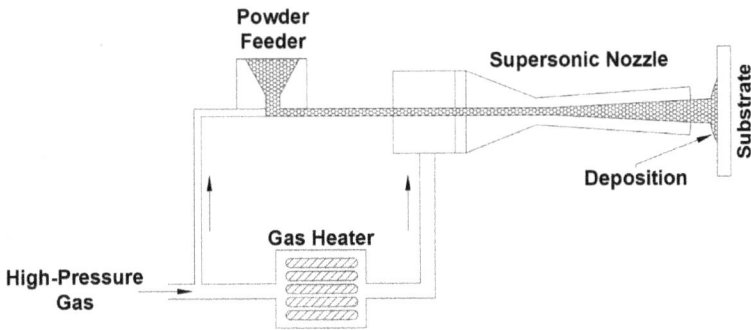

FIGURE 2.3   Schematic of the HPCS process.

The major advantage of LPCS is the relatively low equipment cost. The portability of systems using compressed air allows them to be used in any environment. At the same time, unlike a HPCS system, it can be processed with compressed air which provides very low operating costs. Since particle velocities are lower in LPCS than in HPCS, the stored kinetic energy of the particles is lower. This situation causes the critical speed not to be exceeded, reducing the deposition efficiency. Therefore, particle sizes, morphology, and density of the sprayed powders are more critical in LPCS. The LPCS are more suitable for light and ductile materials, as the deposition efficiency depends on the critical velocity and plastic deformation of the particles that hit the surface (Cavaliere 2018).

2.2.2   HIGH-PRESSURE COLD SPRAY PROCESS

In HPCS systems, particles enter the supersonic nozzle with high-pressure gas. A schematic representation of the HPCS is given in Figure 2.3. In this way, particle velocities in HPCS reach speeds of approximately 800 to 1400 m/s. HPCS makes feasible high-density sprayed particles. N_2 or He gas is preferred as the pressurized gas. Thanks to its high particle velocity, HPCS creates a less porous microstructure than LPCS. In HPCS, a higher pressure powder feed system is required for a mixture of powder. This system increases the cost and causes it to be a larger built-up area. Although the investment and operating cost of the HPCS is high, it can be applied to a wider material group and obtain higher quality deposition layers. Also, it can operate more efficiently due to the high powder feed rate (Cavaliere 2018).

2.2.3   COMPARISON OF COLD SPRAY TECHNOLOGY AND THERMAL SPRAY TECHNOLOGY

The CS method is similar to thermal spray methods in many ways. Pressurized gas is heated to 300–800°C with the help of electrical energy and passes through the converging-diverging nozzle to reach supersonic sonic velocity. Unlike thermal spray methods, the reason for heating the gas is not to melt the powder but to increase the velocity of the gas. Gas passing through the nozzle expands and cools

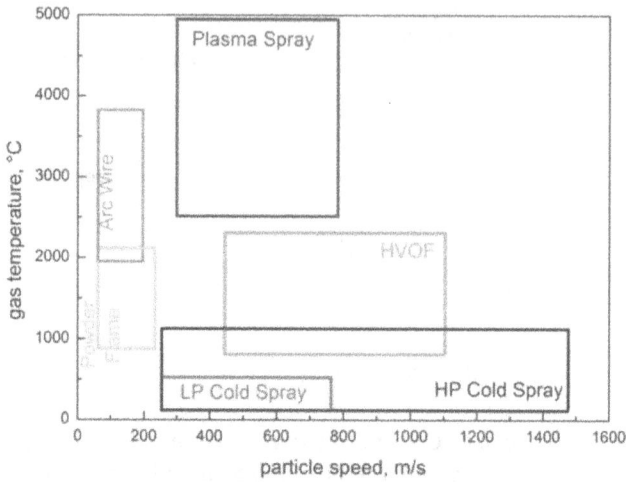

FIGURE 2.4 Gas temperature versus particle speed for thermal spray process (Cavaliere 2015).

rapidly and leaves the nozzle at lower temperatures. The difference between particle velocity and processing temperatures compared to other thermal spray methods is shown in Figure 2.4. Compared to other thermal spray methods, the CS method seems to be operable at very low temperatures and very high particle velocities (Champagne 2007).

Since the operating temperature of traditional thermal spray methods is above the melting temperature of the powders, they have a high tendency to form porosity and oxide. In particular, methods such as atmospheric plasma spray (APS) and high-velocity oxygen fuel (HVOF) contain oxides and pores that reduce the corrosion resistance of the coating (Bala et al. 2013). Also, these oxides and porosities can significantly reduce the mechanical, thermal, and electrical properties of the layers. P. Richer et al. (2010) produced CoNiCrAlY coatings using APS, HVOF, and CS methods in their study. As a result of the EDS analysis of the coating layers, it was reported that the lowest oxide content was obtained in the CS coating thanks to the low processing temperature. In thermal spray coating methods, such as HVOF, operating at high temperatures causes higher residual stress in comparison to CS methods due to the cooling of molten particles (Champagne and Helfritch 2016).

One of the important advantages of the CS method is the deprive of new phase formation thanks to the process performed below the melting temperatures of the particles. In this way, there is no phase transformation between the sprayed powder and the coating layer. The current literature studies show that there is no phase transformation between the initial powder and coating layer by CS (Kılıçay 2020; Padmini et al. 2020). This occasion allows especially critical alloys to be used by the CS method. Tribological properties of different thermal spray methods were also investigated in literature studies. Karaoglanli et al. (2013) investigated the wear properties of CoNiCrAlY coatings using APS, HVOF, and CS methods.

They emphasized that the coating produced by the CS method has the highest hardness and wear resistance properties.

CS coatings also have disadvantages compared to traditional thermal spray methods. The fact that the particles don't melt in this method causes the limitation of the materials to be used. Conventional thermal spray methods can be applied to a wide range of materials, while the CS method is more restricted to ductile materials (Champagne 2007). This is related to the fact that the sprayed material can be plastically deformed when it hits the substrate material. This is more critical in LPCS as the particle velocity is slower. In addition, the amount of gas used in CS processes is higher than in thermal spray methods. Even if this is not a major problem when using air or N_2 gas, it will increase operating costs, especially with the use of He, which is an expensive gas that also provides the highest particle velocity.

2.3 THE EFFECT OF MAIN PROCESS PARAMETERS ON COLD SPRAY

To obtain a good coating layer in CS, the carrier gas, powder feeding unit, and supersonic nozzle system must be processed in harmony. Figure 2.5 shows the basic process parameters affecting the CS coating process.

2.3.1 Nozzle Design

One of the most important parts of a CS coating system related to high deposition efficiency is the supersonic nozzle component. Gas temperature, nozzle movements, trajectory spray distance, and shape of metal powder particles affect nozzle design (Lupoi 2013).

2.3.1.1 Nozzle Traverse Speed

The nozzle traverse speed is a measure of the dwell time of the nozzle on the substrate and the number of metal powder particles deposited per unit time. It also determines the coating thickness deposited on the substrate and affects the

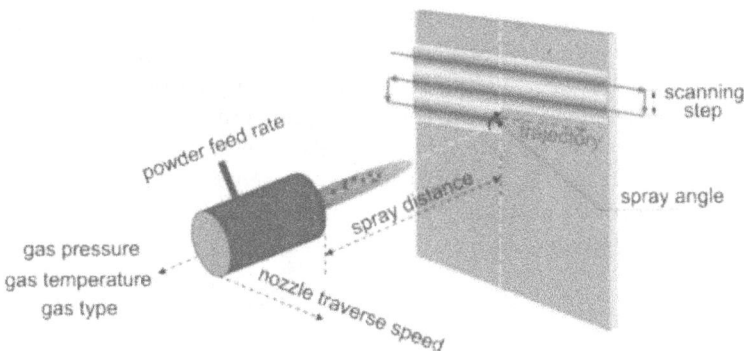

FIGURE 2.5 Schematic representation of CS process parameters (Yin et al. 2018).

cross-section profile (Yin et al. 2018). Lower nozzle traverse speed increases the layer temperature. A denser but softer coating layer is obtained (Wong et al. 2010). If the nozzle travel speed is not high enough for the powder feed rate, the amount of coating deposition increases too much or is segregated in a certain area (Ozdemir et al. 2017). In a CS application, it was observed that the maximum temperature on the coating surface decreased due to the increase in the nozzle traverse speed and, therefore, the decrease in the processing time. It was also noticed that at higher nozzle traverse speed, it takes less time for the coating to reach maximum temperature and thermal equilibrium. In addition, as the nozzle speed gradually increased, the deposition efficiency decreased as a result of the decrease in the substrate-coating temperature (Chen et al. 2017a).

2.3.1.2   Nozzle Scanning Step

The nozzle scanning step is defined as the distance between two consecutive coating passes. It has effects on the coating thickness and the profile of the coating. As the nozzle scanning step distance increases, the maximum-minimum and average coating thickness and flatness of the coating surface decrease (Cai et al. 2013). A nozzle scanning step is generally chosen by researchers to achieve a satisfactory level of flatness on the coating surface. Half of the width of a single coating pass is usually preferable. However, thanks to the recently developed software, the best nozzle scanning step can be performed with simulations to obtain less wavy surfaces (Yin et al. 2018).

2.3.1.3   Gas Temperature

High pressure and high temperature cause the carrier gas to reach high velocities and accelerate the metal powder particles to a higher velocity. Although the gas was initially heated to 400–1100°C, the term "cold spray" was used to describe this process due to the relatively low temperatures (100–500°C) of the expanded gas stream exiting the nozzle (Champagne and Helfritch 2016). The risk of clogging and the thermal resistance of the nozzle limit the temperature used in the system. These obstacles limit the particle impact velocity and impact temperature. The particle impact temperature reduces the critical velocity to be reached due to thermal softening (Schmidt et al. 2006). In addition, in various studies in the academic field, it has been observed that the bond strength increases with the increase of gas temperature, also the microhardness and conductivity properties of the coatings increase as the porosity values decrease (Assadi et al. 2011; Meng et al. 2011; Sudharshan Phani et al. 2007). As a result of numerical and experimental studies, the formula obtained by Suo et al. (2015) revealed that the particle velocity is related to the type of process gas and its temperature (Eq. 2.1). In this relationship, v is the velocity, M is the local Mach number, γ is the specific heat ratio, Mw is the molar mass, T is the local temperature, and R is the ideal gas constant.

$$v = M\sqrt{\frac{\gamma RT}{M_w}} \qquad (2.1)$$

2.3.1.4 Standoff Distance and Spray Angle

The standoff distance, which defines the distance between the exit point of the nozzle and the substrate to be coated, affects the particle velocity and deposition efficiency. In the studies, it was observed that the deposition efficiency decreased with the increase of the standoff distance. There may be different best standoff distance values for the best coating efficiency for different materials and powder groups. When this optimum value is exceeded, the efficiency decreases. In addition, the coating thickness decreases with increasing standoff distance (Li et al. 2008). When the distance between the nozzle and the substrate material is small, a shock wave may occur when the gas jet hits the substrate material. This occasion reduces the particle impact intensity, which deteriorates the deposition quality and efficiency. When the ideal standoff distance is selected, no shock wave occurs, and if the gas velocity is higher than the velocity of the particles, the deposition efficiency increases. If the standoff distance is higher than the ideal, the gas velocity will decrease to a lower level than the particle velocity and the gas will have a slowing effect on the particles (Pattison et al. 2008).

Cai et al. (2013) carried out a mathematical analysis of the coating profile using the numerical solution method in the MATLAB program. The Gaussian distribution was chosen for the simulations because the curve was symmetrical in the first visual analysis. For the coating profile, the relationship in Eq. 2.2 was obtained in accordance with the Gauss equation.

$$Z(x) = \frac{K}{\sigma\sqrt{2\pi}}\exp\left(-\frac{(x-\mu)^2}{2\sigma^2}\right) \tag{2.2}$$

σ is the standard deviation of the Gaussian equation, μ is the mean value of the equation, $Z(x)$ is the height of the coating profile, $S(x)$ is the surface of the coating profile, and K is the constant coefficient of the Gaussian equation. The surface of the 2D coating profile $S(x)$ is given in Eq. 2.3.

$$S(x) = \int_0^1 Z(x)\,dx = \int_0^1 \frac{K}{\sigma\sqrt{2\pi}}\exp\left[-\frac{(x-\mu)^2}{2\sigma^2}\right]dx$$

$$= K * \int_0^1 \frac{K}{\sigma\sqrt{2\pi}}\exp\left[-\frac{(x-\mu)^2}{2\sigma^2}\right]dx \tag{2.3}$$

The surface under the curve of the Gaussian function is equal to 1. Then the surface of the 2D coating profile is given by Eq. 2.4.

$$S(x) = K \tag{2.4}$$

In the experimental studies carried out between 10 mm and 70 mm spray distance values, the maximum $S(x)$ value was obtained at 7.8 mm^2 and 50 mm spray distance. Although the decrease in $S(x)$ value was not more than 2 mm^2, it was observed that there was a decrease when 10 mm and 70 mm spray distances were selected (Cai et al. 2013).

The spray angle is defined as the angle between the nozzle axis and the substrate surface. Generally, the best coating quality and deposition efficiency will occur when the spray angle is 90°. However, the spray angle may change when coating samples with complex geometry or in case of limitations due to operating conditions (Chen et al. 2017b). When the particle hits the substrate, the plastic deformation undergoes till it reaches the normal velocity. For this reason, the normal velocity component of particles ejected at an angle different from the normal angle decreases. This changes the deposition quality and microstructure of the coating (Li et al. 2002). In addition, when the perpendicularity between the nozzle and the substrate changes, the porosity value increases, and the particle bonding and tensile strengths decrease (Binder et al. 2010).

2.3.1.5 Nozzle Trajectory

The studies on the nozzle trajectory, which defines the movements of the nozzle and the path it follows during the CS coating application, affect the quality, homogeneity level, and microstructure properties of the coating, although not much compared to other CS parameters (Chen et al. 2017c; Yin et al. 2018). The nozzle trajectory provides a suitable deposition for the profile of the damaged area to be repaired. Nozzle trajectory strategy controlled by robotic mechanisms and software significantly affects repair quality and accuracy. In repair applications using an appropriate strategy prevents the use of excess material and reduces the process of bringing the part to its original dimensions after processing. Since the damage type and profile of each damaged part will be different, the appropriate nozzle trajectory strategy should be selected so that more efficient and low-cost applications can be made (Wu et al. 2021).

2.3.2 POWDER MORPHOLOGY

Powder particles used in CS expose a plastic deformation due to the impact. When the microstructure images of the cross-sectional coating layer are examined, it is seen that the deformed particles take a form similar to the shape of a blobfish (Figure 2.6), which is flattened due to high pressure in deep seas (Jeandin et al. 2014).

In a study, it was seen that the morphology of the Cu powder particles affects the velocity of the particles in the gas flow. It has been observed that the use of irregular instead of spherical type increases the velocity of the particles in the gas flow (Ning et al. 2007). For cold sprayed Al_2O_3 reinforced Al powder particles, it was determined that the deposition efficiency increased after a certain value with the increase of Al_2O_3 addition. A different result was encountered in the coating process using spherical Al_2O_3 particles. Spherical Al_2O_3 does not form roughness on the coating surface, which provides better deposition efficiency. Therefore, it does not affect the deposition efficiency (Fernandez and Jodoin 2019). This phenomenon was also observed for Ti content coatings. It has been observed that powder spongy and

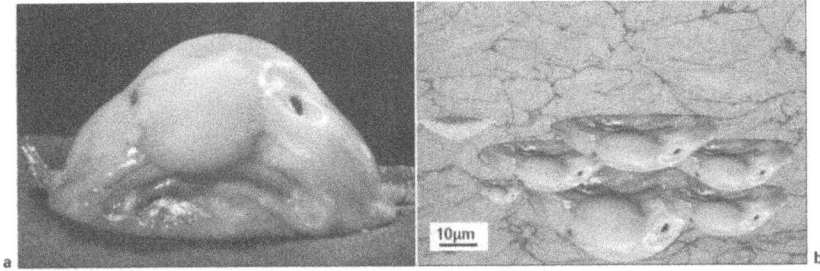

FIGURE 2.6 (a) Blobfish and (b) SEM image of cold sprayed Cu with inserted blobfish (Jeandin et al. 2014).

irregular particles have higher deposition efficiency than spherical ones for pure Ti (Wong et al. 2013). It is known that smaller-sized particles reach higher velocities in the CS process and it is thought that a good deposition efficiency is achieved in this way, although the velocity of small particles when they hit the part can be much lower despite their high exit velocity from the nozzle. The parameter that correlates the particle velocity with the deposition process in the CS process is the critical velocity. The critical velocity is defined as the minimum particle velocity required for a good deposition process. Particle velocities below the critical velocity cause material loss due to erosion and good deposition cannot be achieved (Poirier et al. 2019). According to the results based on simulation and experimental studies in CS coating application of copper (Cu) particles, it was revealed that the critical velocity is a function of particle diameter (dp), and impact temperature (Tp). According to Eq. 2.5, it is concluded that the critical velocity decreases when the size of the particles increases (Schmidt et al. 2009).

$$v_{crit}\left(d_p,T_p\right)=\frac{\sqrt{657000-600*T_p}}{d_p^{0.18}} \tag{2.5}$$

2.4 GREEN APPLICATIONS BY USING COLD SPRAY TECHNOLOGY

With the increase in studies both in the industrial field and in the academic world, the application potential of CS technology has started to be understood more clearly, and its applicability, advantages, and disadvantages have been revealed. Research and studies related to which materials it can be applied to, optimum process parameters, postprocess performance are increasing gradually. Thanks to the repair of metallic-based materials, especially used in aviation, automotive, maritime, and other industrial areas, product lifespan increases. This situation saves the need for remanufacturing material, time, and energy. For this reason, the CS repair method is considered a green method, as it eliminates the harmful effects on the environment. At the same time, the fact that it can be used for additive manufacturing, which is an advanced manufacturing method, makes CS technology more interesting. Along with the production advantages it provides, the cold spraying method also has important advantages such as reducing environmental impacts, economic efficiency, and safety.

2.4.1 Anti-Pathogenic Applications

Interest in the development of materials with antibacterial properties for different biomedical applications is gradually increasing. Carbon steel which has low cost and sufficient mechanical properties widely used in public places and hospitals. However, the formation of various pathogenic biofilms could be observed in such environments (da Silva et al. 2021). CS technology has emerged in recent years as a promising candidate for creating anti-pathogenic coatings. It is possible to coat some bioactive metals such as Ag, Cu, and Zn on different substrate materials by cold spraying (Ghosh et al. 2020). Sanpo and Tharajak (2017) fabricated the Ag, Ni, Zn, and Cu-substituted Hydroxyapatite (HA)/poly-ether-ether-ketone (PEEK) composite coatings on the glass slide substrates by the CS coating technology. They investigated the antibacterial properties of the coatings against *Staphylococcus aureus*. They indicated that cold sprayed coatings have a significant killing effect on *Staphylococcus aureus* when compared with the control group. They determined the bacterial killing rate as HA-Ag/PEEK, HA-Zn/PEEK, HA-Cu/PEEK, and HA-Ni/PEEK, from best to worst, respectively. However, they emphasized that the antibacterial properties of CS coatings should also be investigated with the different types of bacteria. da Silva et al. (2019) investigated the corrosion and antibacterial performance of cold sprayed coatings produced by using Cu powder on the carbon steel substrate. At least 99% purity Cu powder was used. The average grain size was measured as 31 ± 2 µm. XRD analysis confirmed that the sprayed powder and the coating layer were quite similar phases. They determine the antibacterial activity of the Cu coating against *Staphylococcus aureus* using the plate count method. Colony-forming units (CFU) values upon contact time of 0, 5, and 10 minutes are given in Figure 2.7. They determined that after 5 minutes of contact on the Cu coating, the bacteria decreased from 1.97×10^8 to 0.97×10^6 CFU/mL, and after 10 minutes of contact, all *Staphylococcus aureus* bacteria were inactivated. They attributed this situation to the formed Cu ions and moisture at the contact of the coating layer that destroy cell walls of bacteria and prevent their growth. Therefore, they explained that the amount of released Cu ions determines the antibacterial activity. With the obtained results, they emphasized that the cold sprayed Cu coating on the common contact surfaces such as public and hospital environments can be used as an antibacterial surface.

da Silva et al. (2021) coated 99.9% pure Cu powder on low carbon steel by CS method. They investigated the anti-microbial activity and biocompatibility of the produced coating. The cold sprayed coatings were tested against the Methicillin-susceptible *Staphylococcus aureus*, *Escherichia coli*, and *Candida albicans* biofilms for the anti-microbial tests, and biocompatibility of the cold sprayed coatings were investigated in human monocytes. When the cold sprayed Cu surface and the steel surface were compared, they determined that the metabolic activity and growth rates of biofilms decreased significantly (Figure 2.8). At the same time, it was determined that the cold sprayed Cu surfaces had biocompatibility

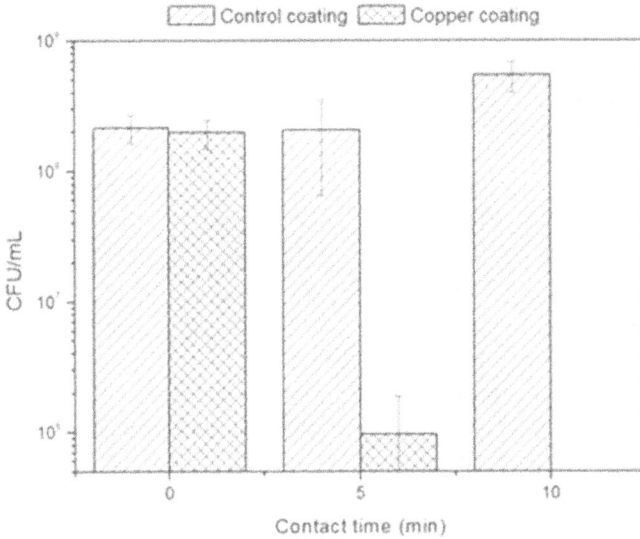

FIGURE 2.7 CFU values of coatings based on contact time (da Silva et al. 2019).

properties. According to their results, they stated that cold sprayed Cu coatings may be interesting to use in healthcare environments due to both anti-microbial and biocompatibility properties.

Hutasoit et al. (2020) produced coatings using Cu powder with a particle size of 5–60 μm with 99.9% purity by CS method on a stainless steel substrate. They investigated the effects of the coatings against the SARS-CoV-2 (COVID-19) virus. They found that Cu-coated surfaces significantly reduced the activation ability of the COVID-19 virus compared to stainless steel. They determined that after a 2-hour incubation period, Cu-coated samples inactivated 96% of the virus. In addition, they

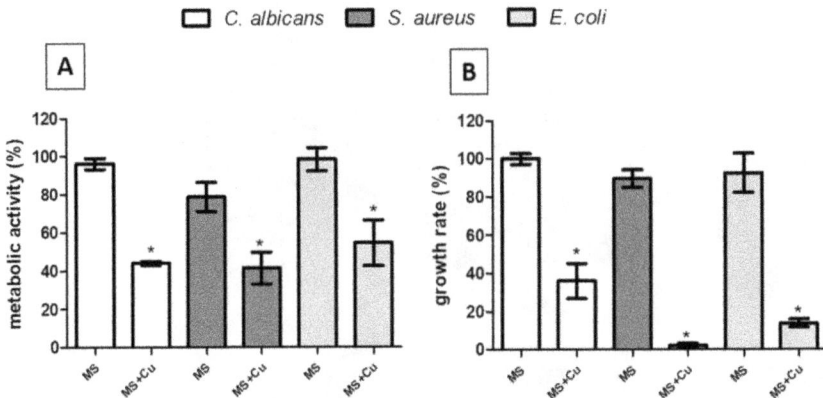

FIGURE 2.8 Metabolic activity and growth rates of biofilms. MS: Metal surface; MS+Cu: Cu coated surface (da Silva et al. 2021).

found that Cu-coated surfaces reduced the lifespan of the COVID-19 virus by more than 5 hours. They emphasized that cold sprayed Cu coatings could easily eliminate viruses in a short time. So one can be implied that cold sprayed Cu content coating can be used on metallic parts in the common use area which can act as an inhibitor for the spread of the virus.

2.4.2 REPAIR APPLICATIONS

Al and Mg alloys are widely used in the aerospace and automotive industry due to their low density and high strength ratio. In addition to this situation, there are various studies on the repair of cold sprayed Al and Mg, steel, and superalloys. To make successful repair applications, the tests and controls are of great importance. Dimensional accuracy should be checked after the applied repair process to structural or non-structural workpieces. In addition, desired mechanical, physical, tribological, and microstructural properties should perform satisfactory results under operating conditions. CS repair processes on Al alloys are widely used and have been extensively researched by researchers. In this section, examples of repair processes using CS technology are given and its performance after the repair process is evaluated.

Lee et al. (2007) repaired a plastic injection mold made of Al 6061-T6 alloy (Figure 2.9), which suffered extensive wear damage as a result of the thermal and

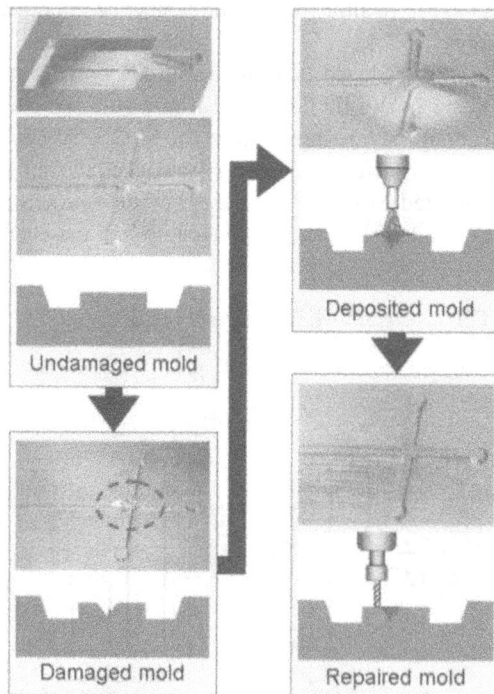

FIGURE 2.9   The schematic representation CS repair process of plastic injection mold (Lee et al. 2007).

mechanical loads. Pure Al powders are used for CS repair technology. As a result of the tests, it was observed that the wear performance of the repaired area improved and its machinability performed similar results with the base material. It was also determined that there was little dimensional difference in the polymer materials produced from the repaired mold and this situation could be eliminated.

Cavaliere and Silvello (2017) repaired the notch damage that occurs in Al2099 alloy plates used in aircraft by using CS. The Al2198 and Al7075 alloy powder particles were used for the repair process. In the crack initiation and growth tests, it was revealed that the repaired plates had six times the crack initiation resistance when the optimum processing parameters were used (Figure 2.10). He attributed this result to the inhibition of crack initiation and propagation thanks to the strong bonding between the repaired area and the substrate. Their results show that damaged Al2029 material is suitable for repairing by CS in terms of fatigue strength.

Widener et al. (2015) repaired flake wear and pitting corrosions on the surface of the sealing hole on an Al 6061 hydraulic valve body used in military navy units with Al 6061 powders (Figure 2.11). It was determined that the porosity value on the repaired surfaces was below 5% and the bond strength was above the required 68.9 MPa. These results show that the repair process was applied successfully.

Shikalov et al. (2017) repaired corrosive damaged plate-shaped aviation parts made of 1163RDTV Al alloy with ASD-1 powder particles. It was observed that the porosity value did not exceed 1% after the repair process, and considering the mechanical properties, it was determined that the elastic area was almost completely equivalent to each other in the stress-strain diagram of the undamaged part and the repaired region.

Cruz et al. (2018) investigated the repair of Al-Cu alloys (Al 2024) and Al-Si alloys (Al C355) used in gearboxes and body parts of aircraft operating under heavy conditions. It has been revealed that the wear behavior of the coatings made on the

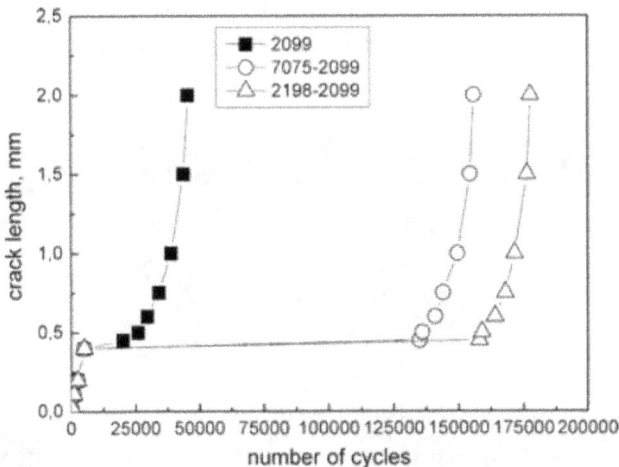

FIGURE 2.10 Diagram of crack length based on the number of cycles repaired alloy Al 2099 using Al 2198 and Al 7075 alloy powders (Cavaliere and Silvello 2017).

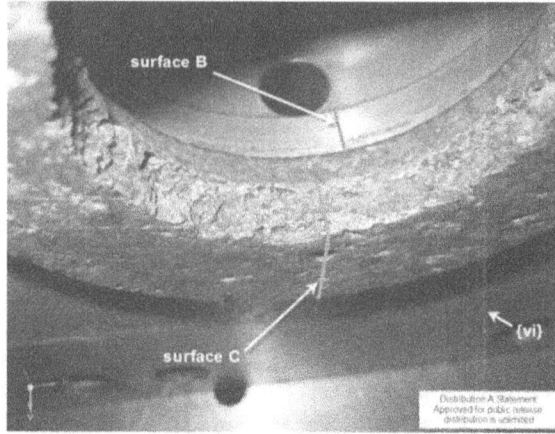

FIGURE 2.11   A macro photo of the damaged and repaired sealing hole surface on an Al 6061 hydraulic valve body (Widener et al. 2015).

same substrate material with the same alloy powders is equivalent. Therefore, they determined that the repair process was successful in terms of tribology.

Astarita et al. (2016) artificially damaged an engine block made of A380 die-cast alloy by machining. The engine block was repaired by two different methods, CS (Figure 2.12) and TIG welding. Microstructure, hardness, and corrosion resistance properties of the repaired zones were investigated. AA4047 alloy was used as filler metal in the TIG welding, and Al-12% Si powder particles were used for CS operation. It was observed that there were no alterations in the microstructure of the substrate material under the coating in the area repaired by CS. They reported that the new surface was found to be quite compact and non-porous. The hardness value of the sprayed coating is close to the substrate material and also shows isotropic properties. It has been stated that the CS method is an alternative to the surface cladding process.

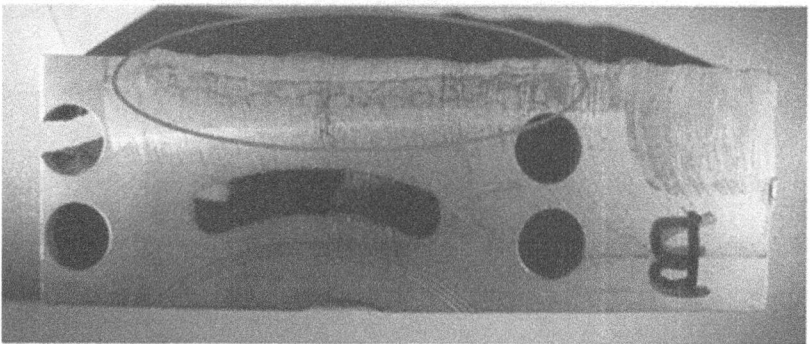

FIGURE 2.12   A close-up photo of engine block repair by CS technology (Astarita et al. 2016).

White et al. (2019) repaired the connection holes of high hardness forged AA7075-T651 and AA2024-T351 plates by CS method. The AA7075-325 and AA2024-325 alloy powders were used. As a result of the experimental studies, it was observed that the CS repair method did not cause any damage to the base material. The micro-hardness of both parts wasn't changed. As a result of the fatigue tests, it has been revealed that the fatigue life is at least equal to the epoxy-based repair applications and even longer life performance in some cases. Although some surface cracks were observed in fractography examinations, it was determined that there was sufficient bonding energy to prevent the separation between coating and substrate material.

In addition to Al alloys, the CS repair method can be applied to various Mg alloys, Ni-based superalloys, and steels. Ogawa and Seo (2011) repaired the part made of Inconel 738LC (IN738LC), which is a Ni-based superalloy and is used in jet engine turbine parts. IN738LC powder particles were used as filler in the process. It has been observed that the use of small particle size positively affects the quality of the coating layer. In small punch tests, He gas content sprayed samples perform higher maximum load-carrying capability. They also applied heat treatment after the CS process. It was seen that the repaired layers were able to recover their crystalline structure and mechanical properties. This situation showed that the adhesion strength of the coating/substrate interface can be improved after heat treatment. The applied heat treatment also ensures a decrease in pores ratio, increase in particles size, and formation of more γ in repaired coating zone. They imply that CS repair of gas turbine blades requires final heat treatment application for better mechanical properties. Faccoli et al. (2014) investigated the repair of damaged ASTM A 743 quality CA6NM cast martensitic stainless steel materials. The materials were repaired by TIG welding and CS method (Figure 2.13) using AISI 316 (austenitic) and AISI 410 (martensitic) stainless steel powder particles. Compressive stresses that occur as a result of repair with CS method improve the fatigue strength properties of the material. The absence of alteration in the microstructure of the repaired zone and the base material interface ensured the preservation of the metallic properties. As a result of the CS method, it was determined that the interface of the surface and the coating base material exhibits higher microhardness than the welding process. They also showed no toughness reduction was observed. According to their results, due to negligible residual stress and extra heat treatment is not required.

Cavaliere et al. (2017) repaired the notched Inconel 718 specimens with the CS method using Inconel 625 particles. They also determined optimum CS repair parameters of superalloy (pressure, temperature, spray distance) concerning porosity reduction and increased adhesion strength and fatigue resistance. Also, aging heat treatments were applied to the repaired samples and it was revealed that it had beneficial effects on fatigue crack resistance and propagation.

The schematic representation of the repair process with the CS process is given in Figure 2.14. A preliminary preparation should be made so that the first layer can adhere to the damaged area. In this step, there is no residue or oxide layer on the surface. A preliminary grinding may be required to make the surface rough (Figure 2.14a). After this step, the damaged area should be filled with suitable powders selected per the base material. The selected powders for filling could be pure, alloy, or mixed.

FIGURE 2.13 Macro photography of CS repair of martensitic stainless steel components (Faccoli et al. 2014).

When a mixed powder is used, the repaired area has a composite microstructure. At the end of the filling process, the part will be larger than its original dimensions. At this step, the repaired surface also becomes quite rough (Figure 2.14b). Machining methods and usually grinding are preferred to provide original dimensions (Figure 2.14c). With the final surface preparation processes, the part is brought to the desired surface quality and tolerance values. Figure 2.15 shows the steps of the repair process with the LPCS system on the damaged Al alloy aircraft wing surface material. The damaged area was successfully filled with densely Al alloy using an N_2 carrier gas. Due to the plastic deformation mechanism caused by the particles hitting the surface, the hardness of the repaired area was determined to be higher than the original coating layer and the filler powder. Residual stress is formed by the

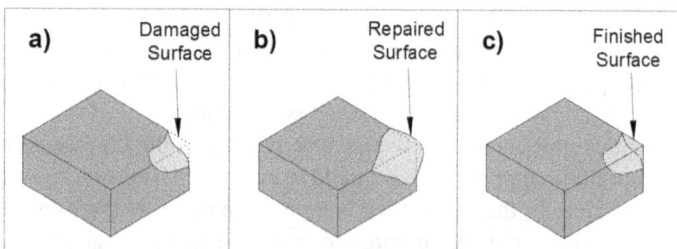

FIGURE 2.14 Schematic representation of part repair with CS technology.

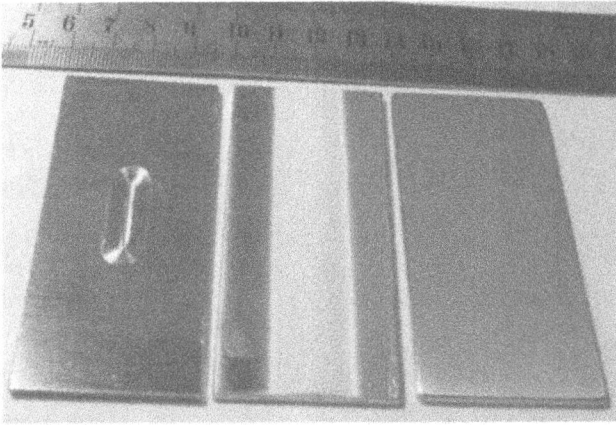

FIGURE 2.15   The repair process of damaged Al part with CS technology (Yandouzi et al. 2014).

CS in the repaired area, which enhances fatigue resistance. At the same time, the repaired zone has a positive effect on corrosion resistance. According to the obtained results in the study, it was determined that the CS process is suitable for the repair of Al alloy aircraft wing surface material (Yandouzi et al. 2014).

Lyalyakin et al. (2015) investigated the repair of damaged bodies of oil pumps of Caterpillar-3116 and Caterpillar-3126 engines by the CS method (Figure 2.16). Sandblasting, CS, machining, and final cleaning steps were performed as repair process steps, respectively. Al_2O_3 powder was used in the sandblasting step to clean the residues in the bodies of the oil pumps. During the CS repair process, pressurized atmosphere gas and Dymet A-80-13 Al powder were used. After the repair process, the pump bodies were machined to their nominal sizes, and Al power and oxides residues were cleaned. After the repairs, they determined that the service times of the oil pump bodies were close to the original pump bodies. They reported that more than 30 oil pump housings were repaired in Moscow between 2012 and 2013. These repaired parts did not cause any malfunction. They found that the cost of repairing the oil pump housing by CS did not exceed about 10–15% of the cost of a new component. They emphasized that the annual economic impact of repair with CS can be cost-effective.

2.4.3   ADDITIVE MANUFACTURING USING COLD SPRAY TECHNOLOGY

With CS technology, powder particles can be accelerated to high speeds to adhere to the substrate. Various structural components can be produced. Therefore, the CS method is also considered an additive manufacturing process (Wang et al. 2015). It facilitates the machinability of small parts. The need for production equipment decreases and production efficiency increases. Since the finished product can be manufactured directly from the raw material, residual raw materials and damage to the environment are eliminated (Pathak and Saha 2017). Metal and composite structures

FIGURE 2.16 Macrophotography of oil pump housings of Caterpillar 3116-3126 engines (a) as damaged (b) as repaired (Lyalyakin et al. 2015).

can be produced with the cold spray additive manufacturing (CSAM) method. With the appropriate powders and process parameters (gas pressure and temperature, powder carrier gas type, nozzle geometry, nozzle movement, speed, etc.), the desired material properties and topologies can be easily produced. Programmed robotic manipulators are generally used to manufacture complex shapes. The microstructural properties such as (pore amount, grain shapes) that affect the product quality and mechanical properties, tribological performances, and residual stresses of the produced parts so it should always be taken into consideration. If necessary, appropriate heat treatments are applied to improve the specified properties. In this section, the structures produced by using different powders with the CSAM method and the evaluation processes of the properties of these structures are examined.

Free-form structural samples produced from Inconel 718 material by CS method and laser melting-based manufacturing method were compared. It is one of the most important advantages of CS method manufacturing is the most homogeneous products, which preserve the microstructure properties of the feed powders in the structure. It was also observed that metallurgical bonding improved and the ductility of the sample increased with various post-production heat treatments. In addition, it has been determined that the fatigue strength of the produced samples can compete with those of casting and forging products. Reduced ductility and size limitations

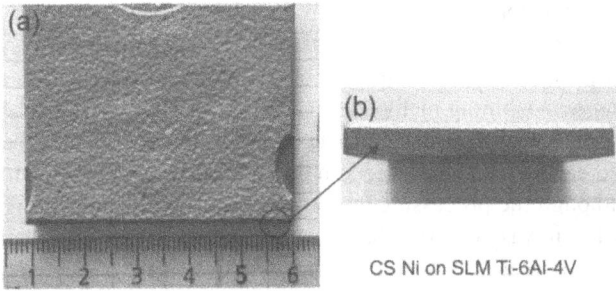

FIGURE 2.17 Ni deposition on Ti-6Al-4V: (a) Surface top view, (b) surface cross-section view (Huang et al. 2018).

due to strain hardening are the issues that need to be developed in the CSAM method (Bagherifard et al. 2018).

Another developed hybrid production method includes CS and laser melting. Ni deposition process was performed on Ti-6Al-4V alloy by using CS and laser melting additive manufacturing methods (Figure 2.17). It has been stated that this method is a potential alternative for joining dissimilar materials. After vacuum heat treatment, it was observed that the unbonded interfaces between the Ni particles improved, while the brittle intermetallic compounds in the structure decreased. However, it has been determined that this process reduces the deposition-substrate bond strength (Huang et al. 2018).

Another technological field in which the CSAM method is used is nuclear technology applications. In a study, a neutron shield was formed by deposition of B_4C/Al (Figure 2.18) on Al 6061-T6 material with the CSAM method. The microstructure, mechanical properties, and neutron shield properties of the material produced under different heat treatment conditions were investigated (200, 300, 400, and 500°C). Maximum ductility and strength and lowest porosity were achieved at 500°C heat treatment conditions. It was observed that the neutron shield performance increased with increasing deposition thickness. In addition, a slight improvement in neutron shield performance was observed for heat treatment at 500°C (Tariq et al. 2018).

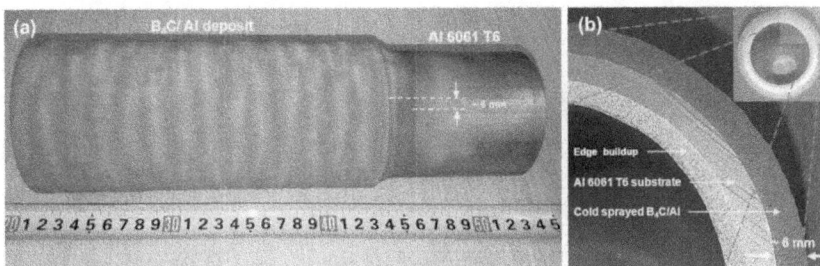

FIGURE 2.18 B_4C/Al deposition on Al 6061 T6: (a) Top view, (b) cross-section view (Tariq et al. 2018).

Cu-containing materials can also be produced with the CSAM method. With the CSAM method, thick and dense Cu coatings were produced on the AA2024 substrate material, and the substrate material was separated from the Cu coating by the electro-discharge machining method after the CSAM method. It has been observed that the performed heat treatments produced a significant effect on the mechanical properties, increasing the tensile strength by 34.2% and reducing the microhardness by 43.6%. Although the processing temperature was below the melting point of the powder particles, it was observed that dynamic and static recrystallization occurred at the interparticle interfaces during CS and heat treatment, respectively. According to the mechanical anisotropy tests, the samples performed different tensile strengths in different directions (Yang et al. 2018).

Xie et al. (2020) produced Al/diamond metal matrix composite structures using the CSAM method with using core-shell diamond powder particles. It was observed that the wear resistance of the produced samples was a superior comparison to laser melted Inconel 625 and 17–4PH alloys. They also showed that the wear performance improved as the number of diamond particles in the composite structure increased.

Chen et al. (2020). produced metal matrix composites were by using the CSAM method. Gas atomized Al7075 alloy powders and Nano-sized TiB_2 powders were used for chemical compositions. The reinforcement phase is dispersed homogenously in the Al7075 matrix. No microstructural defects were observed in the samples. Achieving high particle velocities during production increased the deformation and resulted in improved grain refinement. As a result of grain refinement, better tensile strength results were observed. Also, the addition of TiB_2 increased the hardness of the produced sample. As a result of mechanical tests, fracture morphologies of the samples are brittle. They reported that heat treatment is required for better mechanical results.

Luzin et al. (2020) produced samples using the CSAM method with commercially Ti powders. They investigated the residual stresses. To obtain dense and nonporous Ti structures, optimum process parameters were selected to achieve the highest particle velocity. Ti was deposited as a thick planar coating on the stainless steel substrate and as a thick rod wall on the substrate Al material (Figure 2.19). In both cases, residual stresses were found due to the difference in the thermal expansion coefficient between the substrate and the Ti structure.

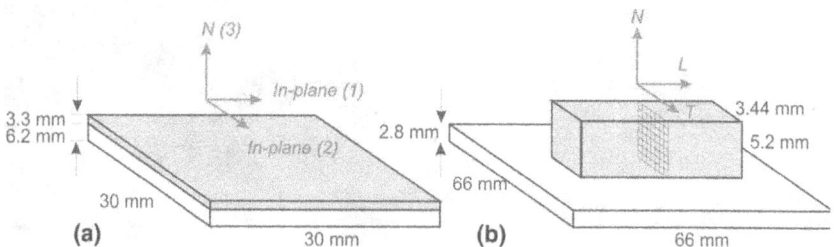

FIGURE 2.19 Two different geometries are produced by CSAM: (a) Thick planar type and (b) thick wall type (Luzin et al. 2020).

To eliminate the residual stress is to choose the same substrate as the deposition powders or to separate the substrate material from the structure after the CSAM method. Another residual stress that occurs in the structure is caused by the deposition stress in the CS process. The control of this residual stress is achieved by adjusting the pressure and temperature parameters in the CS process. In addition, the formed compressive stresses on the surface, which is a characteristic of the CSAM method, have a positive effect on the structure and delay the formation and propagation of cracks.

Another material group that can be produced by the CSAM method is INVAR alloys. Chen et al. (2021) produced Invar 36 alloy structures by the CSAM method. The microstructural properties, pore amount, residual stress, thermal expansion, and mechanical performance were investigated concerning heat-treated and non-heat treated samples. It was observed that the level of porosity decreased with the increase in gas pressure. Phase change and oxidation did not occur in the microstructure during the CSAM process. There are residual stresses with compressive character in the not heat-treated condition. Samples produced with He gas perform higher maximum tensile strength and elongation than Samples produced with N_2 gas. After the heat treatment, the materials produced with both gases perform equivalent yield strengths.

Al 6061 alloy, which is widely used in structural applications, has important advantages due to its high strength/weight ratio. One of the most important factors in the additive manufacturing of materials in the CSAM method is the spray angle. Hutasoit et al. (2021) produced material using gas atomized Al6061 alloy powders deposited on Al5005 alloy substrate with 45° and 90° spray angles (Figure 2.20).

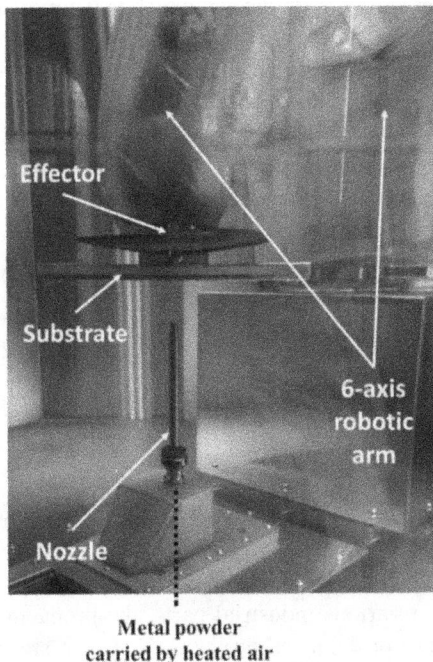

FIGURE 2.20   A representative block production by CSAM method (Hutasoit et al. 2021).

They also investigated the effect of extra heat treatment on the produced material. The untreated samples perform low ductility and low tensile strength while heat-treated samples perform improvement in mechanical properties which are nearly equivalent to forged products. The main reason for this is thought to enhance metallurgical bonding. In addition, they also report that heat treatment increased the corrosion resistance of the material.

2.5 FUTURE PERSPECTIVE

Recently, CS technology has undergone significant development. Currently, CS technology has a wide range of uses such as corrosion, oxidation and wear-resistant metal or alloy coatings, repair of damaged components, additive manufacturing method, and biomedical applications. Currently, studies are carried out in many international research institutions and universities to develop innovative approaches and existing practices. The future of CS technology depends on the potential of innovative materials to be deposited on different substrates with extremely low thermal load and cost. These developments will make the opportunities of CS come to the pioneer.

Improvements in CS equipment are expected, especially with technological developments. For example, He gas is used for high spraying speed, but it is very expensive in terms of cost. Economic efficiency can be increased with gas recycling systems. In addition, technological developments such as lower pressure powder feeding systems and reuse of backscattered powder particles without sticking to the coating surface can be expected. Higher particle velocities and deposition efficiency can be achieved by improving the supersonic nozzle design, which is the most important part of CS systems. In addition, it is known that when the ductile powder is used, even in a short usage time (e.g. 10 minutes), adhesion occurs on the nozzle wall and disrupts the gas flow. Although this situation can be resolved with the use of plastic nozzles, it limits the use of gas temperature and may require more frequent nozzle replacement due to the lower wear resistance of plastics. With technological developments, it will be possible to solve this chronic problem and have longer spray times (Champagne and Helfritch 2014). With technical and economic developments, new application areas can be feasible. It is expected to be used in fields such as wind energy, photovoltaic (photovoltaic) energy, architecture, and medicine. For example, recently, researchers focus on using photovoltaic applications in the production of complex conductive products in solar cells. To improve the surface properties of polymer matrix composites, wind turbine blades CSAM is also utilized (Cavaliere 2018). All aside, CS technology is interesting in anti-pathogenic applications of equipment used in hospitals and dentistry by coating high purity Cu.

CS technology is promising for many applications thanks to its advantages such as reducing oxidation, avoiding undesired phase transformations, having similar initial powder properties in the deposition layer, applying to materials that are sensitive to temperature, and being suitable for a wide range of materials. For this reason, it allows the coating of various industrial parts, the production of additive manufacturing, and the repair of damaged products. However, since the desired surface properties and tolerances cannot be directly achieved in both repair and additive manufacturing processes in current applications, machining may be required after deposition. By developing innovative powder feeding systems and improving nozzle traverse speed in CS, it will be possible to obtain better surface properties and reduce or eliminate secondary processes (Champagne and Helfritch 2014). On the other hand, CS technology provides a clean production opportunity without burning fossil fuels in its current applications. It stands out with its applications such as repairing damaged parts. Low processing temperatures provide low energy consumption and raw material savings. For these reasons, considering the environmental effects, the CS method remains in the category of "green techniques and methods" in current literature and is kept its novelty for future studies.

2.6   CONCLUSIONS

In this section, CS technology is explained in detail and its green applications are examined. CS technology is a thermal spray coating method in which dense coating layers can be produced with a high deposition rate. Deposition layers are produced by the mechanical locking mechanism, which is formed as a result of the impact of the powder particles sprayed from the nozzle at supersonic speeds on the substrate. It is defined as cold since the accelerated sprayed powder particles are solid state due to the process temperature below powder melting temperature. Thanks to the important advantages it provides, this method can also be used in advanced manufacturing methods such as additive manufacturing and innovative methods such as repairing damaged parts. For this reason, it is defined in advanced manufacturing methods and can also be evaluated in the green applications class due to its environmental effects.

It is possible to repair quite different machine parts with CS technology. It is clear that the repair process, instead of replacing the part with a new one, will provide economic and time savings. Thanks to the application of CS at low operating temperatures, it does not create an additional thermal load on the damaged part. Therefore, no microstructural or significant mechanical property alteration is observed in the repaired parts. The service life of the part can be improved by obtaining a hard and wear-resistant depositing layer in the repaired area. In particular, portable low-pressure CS devices also provide mobility for large stationary machine parts. Additional machining operations may be required to obtain the desired surface quality and tolerance values. With the CS technology, it has been possible to produce anti-pathogenic coatings by using bioactive metals such as Ag, Cu, and Zn on different substrate materials. The microstructure of the initial powder and the coating layers is quite similar, and the sprayed powder does not change any phase transformation thanks to the low processing temperatures. Therefore, high purity coatings of bioactive metals such as Ag, Cu, and Zn can be formed by the CS method which these coating layers are quite successful for anti-pathogenic properties.

The use of CS technology as an additive manufacturing method provides a significant cost advantage as it eliminates many manufacturing process steps. The need for equipment such as tools and apparatus for production is reduced. Waste generation and environmental impacts can be minimized, as raw materials can be turned directly into finished products. The produced products may need a small amount of extra surface treatment for the desired surface quality and tolerances.

Therefore, CSAM is classified as a net shape or nearly net shape manufacturing method. Although it can be said that the production of complex-shaped parts needs systemic development.

It can be interpreted that CS technology, which is an advanced manufacturing method, is a very interesting and advanced method for green applications. Also, it is an innovative method that is open to development due to having high potential.

REFERENCES

Ashokkumar, M., D. Thirumalaikumarasamy, P. Thirumal, and R. Barathiraja. 2021. "Influences of Mechanical, Corrosion, Erosion and Tribological Performance of Cold Sprayed Coatings A review." *Materials Today: Proceedings*. doi: 10.1016/j.matpr.2021. 01.664.

Assadi, H., T. Schmidt, H. Richter, J. O. Kliemann, K. Binder, F. Gärtner, T. Klassen, and H. Kreye. 2011. "On Parameter Selection in Cold Spraying." *Journal of Thermal Spray Technology* 20 (6):1161–1176. doi: 10.1007/s11666-011-9662-9.

Astarita, A., F. Coticelli, and U. Prisco. 2016. "Repairing of an Engine Block Through the Cold Gas Dynamic Spray Technology." *Materials Research* 19 (6):1226–1231. doi: 10.1590/1980-5373-mr-2016-0109.

Bagherifard, S., S. Monti, M. V. Zuccoli, M. Riccio, J. Kondás, and M. Guagliano. 2018. "Cold Spray Deposition for Additive Manufacturing of Freeform Structural Components Compared to Selective Laser Melting." *Materials Science and Engineering: A* 721:339–350. doi: 10.1016/j.msea.2018.02.094.

Bala, N., H. Singh, J. Karthikeyan, and S. Prakash. 2013. "Cold Spray Coating Process for Corrosion Protection: A Review." *Surface Engineering* 30 (6):414–421. doi: 10.1179/1743294413y.0000000148.

Binder, K., J. Gottschalk, M. Kollenda, F. Gärtner, and T. Klassen. 2010. "Influence of Impact Angle and Gas Temperature on Mechanical Properties of Titanium Cold Spray Deposits." *Journal of Thermal Spray Technology* 20 (1–2):234–242. doi: 10.1007/s11666-010-9557-1.

Cai, Z., S. Deng, H. Liao, C. Zeng, and G. Montavon. 2013. "The Effect of Spray Distance and Scanning Step on the Coating Thickness Uniformity in Cold Spray Process." *Journal of Thermal Spray Technology* 23 (3):354–362. doi: 10.1007/s11666-013-0002-0.

Cavaliere, P. 2015. "Cold Spray Coating Technology for Metallic Components Repairing." In *Through-life Engineering Services*, 175–184. Springer International Publishing. doi: 10.1007/978-3-319-12111-6_11.

Cavaliere, P. 2018. *Cold-Spray Coatings*. Springer International Publishing.

Cavaliere, P., A. Perrone, and A. Silvello. 2017. "Fatigue Behaviour of Inconel 625 Cold Spray Coatings." *Surface Engineering* 34 (5):380–391. doi: 10.1080/02670844.2017.1371872.

Cavaliere, P., and A. Silvello. 2017. "Crack Repair in Aerospace Aluminum Alloy Panels by Cold Spray." *Journal of Thermal Spray Technology* 26 (4):661–670. doi: 10.1007/s11666-017-0534-9.

Champagne, V. K. 2007. *The Cold Spray Materials Deposition Process Fundamentals and Applications*. Woodhead Publishing.

Champagne, V., and D. Helfritch. 2014. "Critical Assessment 11: Structural Repairs by Cold Spray." *Materials Science and Technology* 31 (6):627–634. doi: 10.1179/1743284714y. 0000000723.

Champagne, V., and D. Helfritch. 2016. "The Unique Abilities of Cold Spray Deposition." *International Materials Reviews* 61 (7):437–455. doi: 10.1080/09506608. 2016.1194948.

Chen, C., Y. Xie, C. Verdy, H. Liao, and S. Deng. 2017a. "Modelling of Coating Thickness Distribution and Its Application in Offline Programming Software." *Surface and Coatings Technology* 318:315–325. doi: 10.1016/j.surfcoat.2016.10.044.

Chen, C., Y. Xie, C. Verdy, R. Huang, H. Liao, Z. Ren, and S. Deng. 2017b. "Numerical Investigation of Transient Coating Build-up and Heat Transfer in Cold Spray." *Surface and Coatings Technology* 326:355–365.

Chen, C., S. Gojon, Y. Xie, S. Yin, C. Verdy, Z. Ren, H. Liao, and S. Deng. 2017c. "A Novel Spiral Trajectory for Damage Component Recovery with Cold Spray." *Surface and Coatings Technology* 309:719–728. doi: 10.1016/j.surfcoat.2016.10.096.

Chen, C., Y. Xie, L. Liu, R. Zhao, X. Jin, S. Li, R. Huang, J. Wang, H. Liao, and Z. Ren. 2021. "Cold Spray Additive Manufacturing of Invar 36 Alloy: Microstructure, Thermal Expansion and Mechanical Properties." *Journal of Materials Science & Technology* 72:39–51. doi: 10.1016/j.jmst.2020.07.038.

Chen, C., Y. Xie, X. Yan, M. Ahmed, R. Lupoi, J. Wang, Z. Ren, H. Liao, and S. Yin. 2020. "Tribological Properties of Al/Diamond Composites Produced by Cold Spray Additive Manufacturing." *Additive Manufacturing* 36. doi: 10.1016/j.addma.2020.101434.

Cruz, D., M. Á. Garrido, Á. Rico, C. J. Múnez, and P. Poza. 2018. "Wear Resistance of Cold Sprayed Al Alloys for Aeronautical Repairs." *Surface Engineering* 35 (4):295–303. doi: 10.1080/02670844.2018.1427318.

da Silva, F. S., N. Cinca, S. Dosta, I. G. Cano, J. M. Guilemany, C. S. A. Caires, A. R. Lima, C. M. Silva, S. L. Oliveira, A. R. L. Caires, and A. V. Benedetti. 2019. "Corrosion Resistance and Antibacterial Properties of Copper Coating Deposited by Cold Gas Spray." *Surface and Coatings Technology* 361:292–301. doi: 10.1016/j.surfcoat.2019.01.029.

da Silva, F. S., A. C. A. de Paula e Silva, P. A. Barbugli, N. Cinca, S. Dosta, I. G. Cano, J. M. Guilemany, C. E. Vergani, and A. V. Benedetti. 2021. "Anti-Biofilm Activity and In Vitro Biocompatibility of Copper Surface Prepared by Cold Gas Spray." *Surface and Coatings Technology* 411. doi: 10.1016/j.surfcoat.2021.126981.

Davis, J. R. 2004. *Handbook of Thermal Spray Technology*. ASM International.

Faccoli, M., G. Cornacchia, D. Maestrini, G. P. Marconi, and R. Roberti. 2014. "Cold Spray Repair of Martensitic Stainless Steel Components." *Journal of Thermal Spray Technology* 23 (8):1270–1280. doi: 10.1007/s11666-014-0129-7.

Fernandez, R., and B. Jodoin. 2019. "Cold Spray Aluminum–Alumina Cermet Coatings: Effect of Alumina Morphology." *Journal of Thermal Spray Technology* 28 (4):737–755. doi: 10.1007/s11666-019-00845-5.

Gärtner, F., T. Stoltenhoff, T. Schmidt, and H. Kreye. 2006. "The Cold Spray Process and Its Potential for Industrial Applications." *Journal of Thermal Spray Technology* 15 (2):223–232. doi: 10.1361/105996306x108110.

Ghosh, M., A. Roy, A. Ghosh, H. Kumar, and G. Saha. 2020. "Antibacterial and antimicrobial coatings on metal substrates by cold spray technique: Present and future perspectives." In *Green Approaches in Medicinal Chemistry for Sustainable Drug Design*, 15–45. Elsevier. doi: 10.1016/B978-0-12-817592-7.00002-2.

Huang, C. J., X. C. Yan, C. Y. Chen, Y. C. Xie, M. Liu, M. Kuang, and H. L. Liao. 2018. "Additive Manufacturing Hybrid Ni/Ti-6Al-4V Structural Component via Selective Laser Melting and Cold Spraying." *Vacuum* 151:275–282. doi: 10.1016/j.vacuum.2018.02.040.

Hutasoit, N., M. A. Javed, R. A. R. Rashid, S. Wade, and S. Palanisamy. 2021. "Effects of Build Orientation and Heat Treatment on Microstructure, Mechanical and Corrosion Properties of Al6061 Aluminium Parts Built by Cold Spray Additive Manufacturing Process." *International Journal of Mechanical Sciences* 204. doi: 10.1016/j.ijmecsci.2021.106526.

Hutasoit, N., B. Kennedy, S. Hamilton, A. Luttick, R. A. Rahman Rashid, and S. Palanisamy. 2020. "Sars-CoV-2 (COVID-19) Inactivation Capability of Copper-Coated Touch Surface Fabricated by Cold-Spray Technology." *Manufacturing Letters* 25:93–97. doi: 10.1016/j.mfglet.2020.08.007.

Jeandin, M., G. Rolland, L. L. Descurninges, and M. H. Berger. 2014. "Which Powders for Cold Spray?" *Surface Engineering* 30 (5):291–298. doi: 10.1179/1743294414y.0000000253.

Karaoglanli, A. C., H. Caliskan, M. S. Gok, A. Erdogan, and A. Turk. 2013. "A Comparative Study of the Microabrasion Wear Behavior of CoNiCrAlY Coatings Fabricated by APS, HVOF, and CGDS Techniques." *Tribology Transactions* 57 (1):11–17. doi: 10.1080/10402004.2013.820372.

Kılıçay, K. 2020. "Development of Protective MMC Coating on TZM Alloy for High Temperature Oxidation Resistance by LPCS." *Surface and Coatings Technology* 393. doi: 10.1016/j.surfcoat.2020.125777.

Lee, J. C., H. J. Kang, W. S. Chu, and S. H. Ahn. 2007. "Repair of Damaged Mold Surface by Cold-Spray Method." *CIRP Annals* 56 (1):577–580. doi: 10.1016/j.cirp.2007.05.138.

Li, C., W-Y. Li, and Y.-Y. Wang. 2002. "Effect of Spray Angle on Deposition Characteristics in Cold Spraying." In Advancing the Science and Applying the Technology. ASM International, Ohio, USA.

Li, W. Y., C. Zhang, X. P. Guo, G. Zhang, H. L. Liao, C. J. Li, and C. Coddet. 2008. "Effect of Standoff Distance on Coating Deposition Characteristics in Cold Spraying." *Materials & Design* 29 (2):297–304. doi: 10.1016/j.matdes.2007.02.005.

Lupoi, R. 2013. "Current Design and Performance of Cold Spray Nozzles: Experimental and Numerical Observations on Deposition Efficiency and Particle Velocity." *Surface Engineering* 30 (5):316–322. doi: 10.1179/1743294413y.0000000214.

Luzin, V., O. Kirstein, S. H. Zahiri, and D. Fraser. 2020. "Residual Stress Buildup in Ti Components Produced by Cold Spray Additive Manufacturing (CSAM)." *Journal of Thermal Spray Technology* 29 (6):1498–1507. doi: 10.1007/s11666-020-01048-z.

Lyalyakin, V. P., A. Yu Kostukov, and V. A. Denisov. 2015. "Special Features of Reconditioning the Housing of a Caterpillar Diesel Oil Pump by Gas-Dynamic Spraying." *Welding International* 30 (1):68–70. doi: 10.1080/09507116.2015.1030152.

Meng, X., J. Zhang, J. Zhao, Y. Liang, and Y. Zhang. 2011. "Influence of Gas Temperature on Microstructure and Properties of Cold Spray 304SS Coating." *Journal of Materials Science & Technology* 27 (9):809–815. doi: 10.1016/s1005-0302(11)60147-3.

Moridi, A., S. M. Hassani-Gangaraj, M. Guagliano, and M. Dao. 2014. "Cold Spray Coating: Review of Material Systems and Future Perspectives." *Surface Engineering* 30 (6):369–395. doi: 10.1179/1743294414y.0000000270.

Ning, X.-J., J.-H. Jang, and H.-J. Kim. 2007. "The Effects of Powder Properties on In-Flight Particle Velocity and Deposition Process during Low Pressure Cold Spray Process." *Applied Surface Science* 253 (18):7449–7455. doi: 10.1016/j.apsusc.2007.03.031.

Ogawa, K., and D. Seo. 2011. "Repair of Turbine Blades Using Cold Spray Technique." In Advances in Gas Turbine Technology. InTech. doi: 10.5772/23623.

Ozdemir, O. C., C. A. Widener, M. J. Carter, and K. W. Johnson. 2017. "Predicting the Effects of Powder Feeding Rates on Particle Impact Conditions and Cold Spray Deposited Coatings." *Journal of Thermal Spray Technology* 26 (7):1598–1615. doi: 10.1007/s11666-017-0611-0.

Padmini, B. V., M. Mathapati, H. B. Niranjan, P. Sampathkumaran, S. Seetharamu, M. R. Ramesh, and N. Mohan. 2020. "High Temperature Tribological Studies of Cold Sprayed Nickel Based Alloy on Low Carbon Steels." *Materials Today: Proceedings* 27:1951–1958. doi: 10.1016/j.matpr.2019.09.025.

Pathak, S., and G. Saha. 2017. "Development of Sustainable Cold Spray Coatings and 3D Additive Manufacturing Components for Repair/Manufacturing Applications: A Critical Review." *Coatings* 7 (8). doi: 10.3390/coatings7080122.

Pattison, J., S. Celotto, A. Khan, and W. O'Neill. 2008. "Standoff Distance and Bow Shock Phenomena in the Cold Spray Process." *Surface and Coatings Technology* 202 (8):1443–1454. doi: 10.1016/j.surfcoat.2007.06.065.

Poirier, D., J.-G. Legoux, P. Vo, B. Blais, J. D. Giallonardo, and P. G. Keech. 2019. "Powder Development and Qualification for High-Performance Cold Spray Copper Coatings on Steel Substrates." *Journal of Thermal Spray Technology* 28 (3):444–459.

Raoelison, R. N., C. Verdy, and H. Liao. 2017. "Cold Gas Dynamic Spray Additive Manufacturing Today: Deposit Possibilities, Technological Solutions and Viable Applications." *Materials & Design* 133:266–287. doi: 10.1016/j.matdes.2017.07.067.

Richer, P., M. Yandouzi, L. Beauvais, and B. Jodoin. 2010. "Oxidation Behaviour of CoNiCrAlY Bond Coats Produced by Plasma, HVOF and Cold Gas Dynamic Spraying." *Surface and Coatings Technology* 204 (24):3962–3974. doi: 10.1016/j.surfcoat.2010.03.043.

Sanpo, N., and J. Tharajak. 2017. "Antimicrobial Property of Cold-Sprayed Transition Metals-Substituted Hydroxyapatite/PEEK Coating." *Applied Mechanics and Materials* 866:77–80. doi: 10.4028/www.scientific.net/AMM.866.77.

Schmidt, T., H. Assadi, F. Gärtner, H. Richter, T. Stoltenhoff, H. Kreye, and T. Klassen. 2009. "From Particle Acceleration to Impact and Bonding in Cold Spraying." *Journal of Thermal Spray Technology* 18 (5–6).

Schmidt, T., F. Gaertner, and H. Kreye. 2006. "New Developments in Cold Spray Based on Higher Gas and Particle Temperatures." *Journal of Thermal Spray Technology* 15 (4):488–494. doi: 10.1361/105996306x147144.

Shikalov, V. S., N. S. Ryashin, and A. V. Lapaev. 2017. "Cold Spray Repairing Corrosively Damaged Areas on Aircraft Constructions." *Solid State Phenomena* 265:325–330. doi: 10.4028/www.scientific.net/SSP.265.325.

Shushpanov, A. P. A., A. N. Papyrin, V. F. Kosarev, N. I. Nesterovich, M. M. Shushpanov. 1994. Gas-dynamic spraying method for applying a coating. Edited by U.S. Patent and Trademark Office. United States.

Sudharshan Phani, P., D. Srinivasa Rao, S. V. Joshi, and G. Sundararajan. 2007. "Effect of Process Parameters and Heat Treatments on Properties of Cold Sprayed Copper Coatings." *Journal of Thermal Spray Technology* 16 (3):425–434. doi: 10.1007/s11666-007-9048-1.

Suo, X., S. Yin, M.-P. Planche, T. Liu, and H. Liao. 2015. "Strong Effect of Carrier Gas Species on Particle Velocity during Cold Spray Processes." *Surface and Coatings Technology* 268:90–93.

Tariq, N. H., L. Gyansah, J. Q. Wang, X. Qiu, B. Feng, M. T. Siddique, and T. Y. Xiong. 2018. "Cold Spray Additive Manufacturing: A Viable Strategy to Fabricate Thick B4C/Al Composite Coatings for Neutron Shielding Applications." *Surface and Coatings Technology* 339:224–236. doi: 10.1016/j.surfcoat.2018.02.007.

Wang, X., F. Feng, M. A. Klecka, M. D. Mordasky, J. K. Garofano, T. El-Wardany, A. Nardi, and V. K. Champagne. 2015. "Characterization and Modeling of the Bonding Process in Cold Spray Additive Manufacturing." *Additive Manufacturing* 8:149–162. doi: 10.1016/j.addma.2015.03.006.

White, B., W. Story, L. Brewer, and J. B. Jordon. 2019. "Fatigue Behaviour of Fastener Holes in High-Strength Aluminium Plates Repaired by Cold Spray Deposition." *Fatigue & Fracture of Engineering Materials & Structures* 43 (2):317–329. doi: 10.1111/ffe.13144.

Widener, C. A., M. J. Carter, O. C. Ozdemir, R. H. Hrabe, B. Hoiland, T. E. Stamey, V. K. Champagne, and T. J. Eden. 2015. "Application of High-Pressure Cold Spray for an Internal Bore Repair of a Navy Valve Actuator." *Journal of Thermal Spray Technology* 25 (1–2):193–201. doi: 10.1007/s11666-015-0366-4.

Widener, C. A., O. C. Ozdemir, and M. Carter. 2018. "Structural Repair Using Cold Spray Technology for Enhanced Sustainability of High Value Assets." *15th Global Conference on Sustainable Manufacturing* 21:361–368. doi: 10.1016/j.promfg.2018.02.132.

Wong, W., E. Irissou, A. N. Ryabinin, J.-G. Legoux, and S. Yue. 2010. "Influence of Helium and Nitrogen Gases on the Properties of Cold Gas Dynamic Sprayed Pure Titanium Coatings." *Journal of Thermal Spray Technology* 20 (1–2):213–226. doi: 10.1007/s11666-010-9568-y.

Wong, W., P. Vo, E. Irissou, A. N. Ryabinin, J. G. Legoux, and S. Yue. 2013. "Effect of Particle Morphology and Size Distribution on Cold-Sprayed Pure Titanium Coatings." *Journal of Thermal Spray Technology* 22 (7):1140–1153. doi: 10.1007/s11666-013-9951-6.

Wu, H., S. Liu, X. Xie, Y. Zhang, H. Liao, and S. Deng. 2021. "A Framework for a Knowledge Based Cold Spray Repairing System." *Journal of Intelligent Manufacturing*. doi: 10.1007/s10845-021-01770-7.

Xie, X., Y. Ma, C. Chen, G. Ji, C. Verdy, H. Wu, Z. Chen, S. Yuan, B. Normand, S. Yin, and H. Liao. 2020. "Cold Spray Additive Manufacturing of Metal Matrix Composites (MMCs) Using a Novel Nano-TiB$_2$-Reinforced 7075Al Powder." *Journal of Alloys and Compounds* 819. doi: 10.1016/j.jallcom.2019.152962.

Yandouzi, M., S. Gaydos, D. Guo, R. Ghelichi, and B. Jodoin. 2014. "Aircraft Skin Restoration and Evaluation." *Journal of Thermal Spray Technology* 23 (8):1281–1290. doi: 10.1007/s11666-014-0130-1.

Yang, K., W. Li, X. Guo, X. Yang, and Y. Xu. 2018. "Characterizations and Anisotropy of Cold-Spraying Additive-Manufactured Copper Bulk." *Journal of Materials Science & Technology* 34 (9):1570–1579. doi: 10.1016/j.jmst.2018.01.002.

Yin, S., P. Cavaliere, B. Aldwell, R. Jenkins, H. Liao, W. Li, and R. Lupoi. 2018. "Cold Spray Additive Manufacturing and Repair: Fundamentals and Applications." *Additive Manufacturing* 21:628–650. doi: 10.1016/j.addma.2018.04.017.

Yu, T., M. Chen, and Z. Wu. 2021. "Experimental and Numerical Study of Deposition Mechanisms for Cold Spray Additive Manufacturing Process." *Chinese Journal of Aeronautics*. doi: 10.1016/j.cja.2021.02.002.

Zhao, Z. B., B. A. Gillispie, and J. R. Smith. 2006. "Coating Deposition by the Kinetic Spray Process." *Surface and Coatings Technology* 200 (16–17):4746–4754. doi: 10.1016/j.surfcoat.2005.04.033.

3 Multi-Criteria Decision-Making Applications in Conventional and Unconventional Machining Techniques

Şenol Bayraktar and Erhan Şentürk
Recep Tayyip Erdoğan University, Rize, Turkey

CONTENTS

3.1 Introduction to MCDM...57
3.2 Why Is the MCDM Preferred? 58
3.3 MCDM Approaches and Their Applications...60
    3.3.1 VIsekriterijumska Optimizacija KOmpromisno
        Resenje (VIKOR)..60
    3.3.2 COmplex PRoportional ASsessment (COPRAS) 63
    3.3.3 MOORA Plus Full Multiplicative Form (MULTIMOORA)............66
    3.3.4 Weighted Aggregated Sum Product ASsessment (WASPAS)...........69
    3.3.5 EVAluation of MIXed Data (EVAMIX) 70
    3.3.6 OCcupational Repetitive Actions (OCRA)....................................72
    3.3.7 Multi-Attributive Border Approximation Area
        Comparison (MABAC) 75
3.4 Concluding Remarks 77
References...78

3.1 INTRODUCTION TO MCDM

Decision-making is an idea or action that is obtained as a result of mental activities among different alternatives throughout life. It constitutes a phenomenon with different orientations such as "right-wrong", "good-bad" or "effective-ineffective". When we look at the existing process of human beings, the results of conscious and unconscious decisions show themselves in all areas. What path should we follow when making decisions in order to reach the most appropriate result? In practice, methods that can meet the expectations of the decision-maker over time are used to create a less inaccurate and realistic decision output. In addition, various methods that involve different preferences and produce different results for

DOI: 10.1201/9780367822385-3

each decision-maker are constantly being developed. Optimum hybrid decision-making methods emerge by using different combinations together. Mathematical and computational methods are used to support the evaluation process of complex criteria with multiple alternatives in many branches of science. Multi Criteria Decision Making (MCDM), with these methods, is a model and analysis supporter that enables the use of qualitative and quantitative factors as well as ranking, comparison and selection criteria. In other words, it creates a series of decisions that are compatible with reason and logic rather than making a decision. The use of MCDM is increasing rapidly in industries such as manufacturing, energy, aerospace, automotive, construction and defense, due to the variety and competitiveness of decision mechanisms. It is also preferred to assist in many different areas such as material selection, procurement, human resources, infrastructure, agriculture, food, health, risk management and occupational safety. Different results can be obtained with different MCDM methods on the same problems in any field. This reveals a limiting feature of MCDM. However, ideal results can be achieved with different combinations or hybrid techniques. MCDM methods are divided into two groups as multiple attribute decision-making (MADM) and multi-objective decision-making (MODM) due to different problem variables. While MADM deals with decision problems with a finite number of alternatives and features, MODM deals with decision problems with an infinite number of alternatives and features. Many multi-objective problems do not have well-defined alternatives. Therefore, different methods are applied according to the nature of the decision problems [1]. Since MCDM is a unique computational technique in which alternatives are ranked, the same results may not be obtained with the same input values and different MCDM techniques. Many MCDM methods can be arranged according to different parameters and provide convenience to solve problems. The use of AHP, SWARA, COPRAS, MULTIMOORA, VIKOR, PSO, TOPSIS, WASPAS, EVAMIX, OCRA, MABAC and hybrid methods can be used to find the best alternative and reach the determined target. Today, it is possible for most decision-makers to obtain more efficient and reliable results due to the use of hybrid methods. In the literature, studies on the effectiveness of hybrid techniques in different sectoral applications and the development of these techniques continue. Studies on the effectiveness of hybrid techniques in different sectoral applications and the development of these techniques continue [1]. In addition, ANN, GRA, FL, GA, PSO, PROMETHEE, AHP and ELECTRE techniques are detailed in [2]. In this book chapter, VIKOR, COPRAS, MULTIMOORA, WASPAS, EVAMIX, OCRA and MABAC techniques that are still being developed, and current studies on processing in the literature related to these techniques are presented in detail and comparatively.

3.2 WHY IS THE MCDM PREFERRED?

MCDM is the method encompassing mathematical and computational tools to support the subjective evaluation of performance criterion. It is a collection of methodologies used to compare, select or rank alternatives involving multiple and conflicting criteria and both tangible and intangible factors. It is used by decision-makers in solving real-world decision-making problems according to their own preferences in order to choose the appropriate one [3]. It has been developed in recent years to cope with the decision-making limitations of human, which arises in complex decision environments, and is widely preferred in many areas. This leads to a tendency to simplify the problem by using heuristic approaches rather than rational or analytical approaches. It can also lead to subjective judgments and loss of important information. It has been determined that experts have difficulties in evaluating complex structures and applying simplified decision-making methods. Thus, better and alternative solutions can be used with MCDM techniques that can cope with large amounts of information and computation [4]. Competition emerges in many sectors depending on the developing conditions. For this reason, fast and reliable decision-making in an increasingly competitive environment comes to the fore. All the tools that can help to choose among the many alternatives are critical. In this context, thanks to the MCDM methods that continue to be developed, the efficiency and effectiveness of these processes can be increased significantly. Decision-making problems, with their powerful tools, can be easily analyzed in complex problems in different fields [1]. In other words, it describes a set of techniques used to combine different evaluation indicators into a general index in ordering alternatives from best to worst (Figure 3.1).

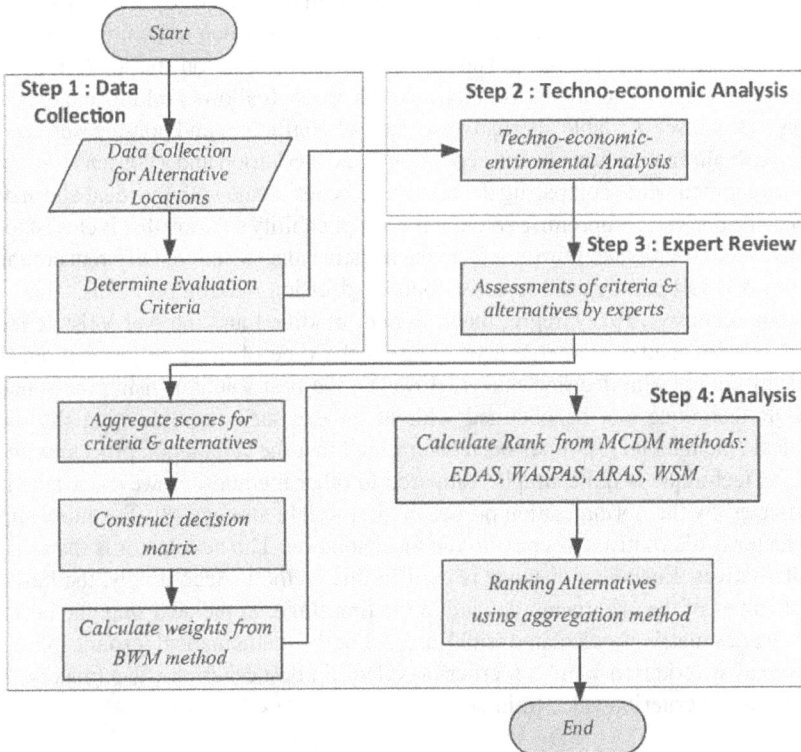

FIGURE 3.1 Example application stages for MCDM techniques [5].

MCDM methods in which a specific approach is used to manage data related to each problem reveal the value function according to Eq. 3.1.

$$f_i : [0,1]^n \rightarrow f_i \in [0,1] \tag{3.1}$$

Here, the value function links the qualitative and quantitative dataset x_j, $j = 1, 2, \cdots, n$ of the criterion vectors to a single numerical value for each alternative i [6]. Although each data set is measurable at its own scale, it is not compatible with other scales.

3.3 MCDM APPROACHES AND THEIR APPLICATIONS

3.3.1 VISEKRITERIJUMSKA OPTIMIZACIJA KOMPROMISNO RESENJE (VIKOR)

VIKOR was first revealed by Opricovic to solve decision problems non-proportional and with contradictory criteria. The closeness of each alternative to the ideal solution is evaluated, and then it is aimed to find a compromise solution with the multi-criteria ranking index in this method. In other words, the aim is to determine an applicability compromise solution that is closest to the ideal. The proposed solution may provide the best compromise due to the contradictory nature of the criteria. It is widely used in computer, engineering, business and resource management [6]. This method has started to be preferred among decision-making techniques due to its simple and easy-to-understand calculation steps. The basic concept is based on the identification of positive and negative ideal points in the solution space. It allows ranking and choosing among a finite set of viable alternatives in case of conflicting and non-measurable criteria. Each alternative is evaluated according to each criterion and a consensus ranking can be obtained while comparing the relative closeness scale with the ideal alternative. Hence, the derived compromise solution is an applicability solution that is closest to the positive ideal solution and furthest from the negative ideal solution. Different combinations of VIKOR can be used to solve complex decision-making problems. These are the comprehensive, fuzzy, regret theory-based, modified and interval VIKOR methods. Different methods emerge depending on the type of decision problem and the needs of the relevant decision-maker. However, the best solutions using the standard VIKOR technique can be obtained without unnecessarily complicating the mathematical calculations [7]. It has been determined that the calculation process with the VIKOR technique is quite simple compared to other methods. However, results may be affected by the normalization procedure and weight strategy [8]. The most important factor is alternative closeness to the ideal solution. The next factor is the ranking of alternatives. Euclidean distance is used in this method. Accordingly, the best and worst values of the criteria are defined in the first stage. In the next step, the best and worst values matrix is calculated with Eq. 3.2. For the mathematical formula in Eq. 3.2 is given as w: criterion weight, f: criterion value, f^*: best criterion value (max or min) and f^-: worst criterion value (min or max).

$$S = w \frac{f^* - f}{f^* - f^-} \tag{3.2}$$

Then, according to Eq. 3.3, S, R and Q values are calculated for each alternative. There is no normalized weight matrix as in TOPSIS, MULTIMOORA and COPRAS techniques [9].

$$Q = v\frac{S_j - S^*}{S^- - S^*} + (1 - v)\frac{R_j - R^*}{R^- - R^*},$$ (3.3)

where v is the decision factor (if $v \geq 0.5$ great compromise, if $v = 0.5$ consent-based compromise, if $v \leq 0.5$ vetoed compromise) according to Eq. 3.3.

$$S_j = \Sigma w\frac{f^* - f_i}{f^* - f^-}, \ S^* = \min S_j, \ S^- = \max S_j;$$

$$R_j = \max\left[w\frac{f^* - f_i}{f^* - f^-}\right], \ R^* = \min R_j, \ R^- = \max R_j$$

The VIKOR method is widely preferred in the literature in studies on machinability. Samson et al. examined roundness, taper angle (T_a), material removal rate (MRR) and surface roughness (SR) using different pressure (140, 160 and 180 MPa), stand-off stance (SOD) (2, 3 and 4 mm) and abrasive flow rate (0.22, 0.32 and 0.42 kg/min) in Abrasive water jet machining (AWJM) process of Inconel 718 material. It was determined that the MRR decreased with the increase in SOD and the optimum experimental parameters were pressure: 180 MPa, abrasive flow rate: 0.42 kg/min and SOD: 2 mm with the VIKOR method. It was stated that roundness: 0.085 mm, MRR: 0.222 gm/s, T_a: 0.098° and SR: 1.365 μm were measured in these parameters and VIKOR technique was successfully used for optimum parameter combination [10]. Singaravel et al. optimized the experimental inputs using the Taguchi-VIKOR hybrid approach for machining AISI D2 die steel using Electrical discharge machining (EDM). While different dielectric fluid (Kerosene, jatropha oil and cottonseed oil), current (6, 8 and 10 A) and pulse on time (200, 300 and 400 μs) were used as independent variables, the experimental outputs were measured as MRR, TWR (Tool wear rate) and SR. Optimum parameters were determined as cottonseed oil, current: 8A and pulse on time: 400 μs with statistical technique. It was stated that TWR decreases as a result of the cryogenic treatment applied to the electrode and the proposed approaches were simple, useful and reliable for the optimization of process parameters [11]. Dey et al. optimized the experimental outputs in the machining of Al6061/cenosphere with the EDM process using a TOPSIS-VIKOR-based approach. They used peak current (PC), pulse on time (T_{on}), percentage of reinforcement (PoR) and flushing pressure (FP) as independent variables. The weighting factors for the criteria were determined by AHP. Optimal parameters for experimental outputs were calculated as PC: 10 A, T_{on}: 1010 μs, PoR: 2% and FP: 0.6 MPa [12]. Khan and Maity optimized the independent variables with the fuzzy-VIKOR approach in turning of commercially pure titanium (CP-Ti) grade 2 workpieces. While different cutting speeds (40, 70 and 100 m/min), feeds (0.05, 0.1 and 0.15 mm/rev) and depths of cut (0.2, 0.4 and 0.6 mm) were used as independent variables, test results were measured as cutting force, tool wear and SR. It was determined that the best

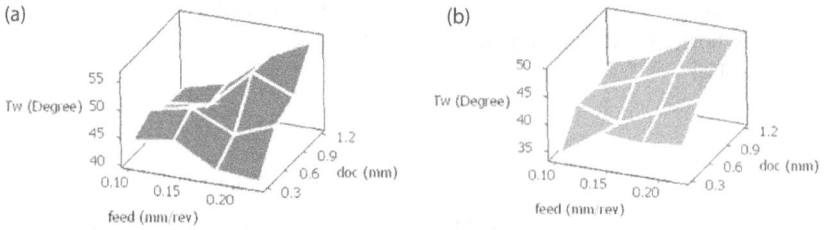

FIGURE 3.2 Workpiece temperature graphs: (a) Dry machining and (b) Mqcl machining [15].

parametric combination was obtained as cutting speed: 40 m/min, feed: 0.05 mm/rev and depth of cut: 0.2 mm. It was found that the proposed approach was systematic and easy to understand. In addition, it was demonstrated that the fuzzy-VIKOR approach could be adopted to obtain the best parametric combination [13]. Kumar et al. optimized TWR, SR and MRR with GRA and entropy-integrated-gray-VIKOR methods using Cu, CuW and graphite tool in the EDM process of Zircaloy-2. It was determined that the maximum MRR was obtained with the negative polarity graphite tool and increased with the increase of T_{on}. It was observed that minimum TWR and SR were obtained with positive polarity using the graphite tool and increased with the increase of T_{on} and PC (Ip). It was revealed that the GRA and entropy-integrated gray VIKOR methods confirmed the 16th alternative as the optimal value, and the entropy-integrated gray VIKOR method was more effective than the methods in the current study [14]. Vikram et al. optimized the machinability properties of AISI 316L using the VIKOR technique under dry, minimum quantity cooling and lubrication (MQCL), different cutting speeds (122, 141, 160 and 179 m/min), feeds (0.10, 0.14, 0.18 and 0.22 mm/rev), depths of cut (0.3, 0.6, 0.9 and 1.2 mm) conditions. Average SR, workpiece temperature (Tw) and tool temperature (TT) were obtained as experimental outputs. The optimum values for minimum SR: 0.72 μm, Tw: 44.1°C and TT: 66°C in dry machining conditions were determined as cutting speed: 122 m/min, feed rate: 0.1 mm/rev and depth of cut: 0.3 mm. Optimum values for minimum SR: 0.87 μm, Tw: 39.1°C and TT: 50°C in MQCL machining conditions were determined as cutting speed: 141 m/min, feed rate: 0.14 mm/rev and depth of cut: 0.3 mm. In addition, discontinuous small helical chips were observed at low feed and depth of cut values (Figures 3.2–3.4) [15].

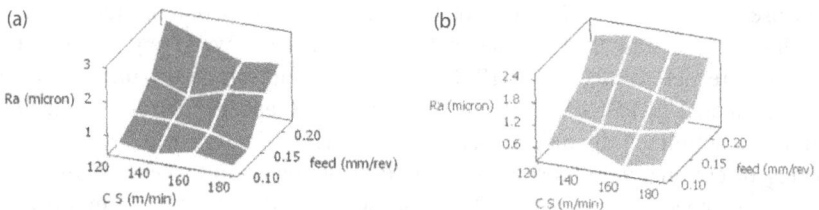

FIGURE 3.3 Surface roughness graphs: (a) Dry machining and (b) Mqcl machining [15].

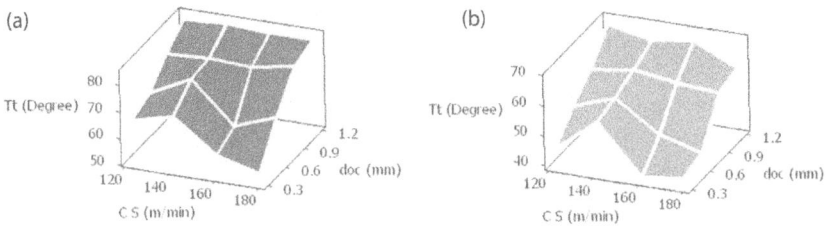

FIGURE 3.4   Tool temperature graphs: (a) Dry machining and (b) Mqcl machining [15].

Phan and Muthuramalingam optimized the SR, MRR, microhardness (HV) and white layer thickness (WLT) in the vibration aided EDM process of high silicon-carbon tool steel using different MCDM techniques. Deng's approach, preference selection index (PSI), COPRAS, GRA, simple-additive-weighting (SAW), TOPSIS and VIKOR techniques were used as MCDM techniques. It was determined that the TOPSIS method provided better estimation accuracy and the frequency was effective in determining the quality criteria based on the TOPSIS method. Optimum parameters measured as I: 4A, T_{on}: 12 µs, T_{off}: 5.5 µs and frequency: 512 Hz [16]. Babu and Jeyapaul optimized the machining parameters by using the Taguchi-VIKOR hybrid approach in EDM process of A6082/Fly Ash/Al_2O_3 hybrid metal matrix composite. While wire type (brass and zinc-coated brass), different T_{on} (108, 110 and 112 µs), T_{off} (56, 58 and 60 µs) and servo feed (1030, 1050 and 1070 mm/min) were used as independent variables, the machining performances were determined as cutting speed (CS), kerf width (KW), wire wear ratio (WWR) and overcut (OC). While the entropy method was used to evaluate the weight of each output, VIKOR was used for the final ranking of the alternatives. Optimum values for maximum CS and minimum WWR, KW and OC were calculated as 3.41 mm/min, 0.07 mm, 0.27 mm and 0.014 mm, respectively [17]. Sahu et al. optimized MRR, TWR and SR with simple optimization (SOPT) and multi-objective simple optimization (MOSOPT) in the EDM process of Ti6Al4V alloy and ranked the alternatives with the VIKOR technique. Different tool type (Cu, $Cu_{90}W_5(B_4C)_5$, $Cu_{80}W_{10}(B_4C)_{10}$ and $Cu_{70}W_{15}(B_4C)_{15}$) current (4, 6, 8 and 10A), T_{on} (50, 100, 150 and 200 µs) and duty cycle (40, 50, 60 and 70%) were used as input variables. It was revealed that while current affected MRR, TWR and tool type, respectively, SR was affected by tool type and current, respectively. It was observed that T_{on} and duty cycle had less effect on performance measures. It was stated that SOPT could be performed parameter optimization with less computational processes and could be easily adapted to multiple optimization problems. The best solution among all Pareto-optimal solutions with the lowest VIKOR index was chosen [18].

3.3.2   COMPLEX PROPORTIONAL ASSESSMENT (COPRAS)

COPRAS covers qualitative and quantitative features and allows the selection of alternatives among the results. The solution is determined by the ratio to the ideal solution. Gradual ranking and degree of benefit using this method can be calculated in case of conflicting criteria. Alternatives must be ranked in descending order in order to obtain alternative rankings taking into account the degree of benefit [8]. It is also preferred to convert a multi-objective review into a single-objective review [19]. Experimental outputs in the first stage using the COPRAS method are collected in the decision matrix according to Eq. 3.4. Here, n is the number of experiments and m is the number of output responses.

$$X = \begin{bmatrix} x_{11} & x_{12} & \cdots & x_{1m} \\ x_{21} & x_{22} & \cdots & x_{2m} \\ \cdots & \cdots & \cdots & \cdots \\ x_{n1} & x_{n2} & \cdots & x_{nm} \end{bmatrix} \tag{3.4}$$

According to Eq. 3.5 each value in the decision matrix is normalized in the second step. Here, x_{ij} is the criterion value and $\sum x_{ij}$ is the sum of the criterion values.

$$\overline{x_{ij}} = \frac{x_{ij}}{\sum x_{ij}} \tag{3.5}$$

The weights of the output responses are determined and according to Eq. 3.6 normalized weighted matrix is calculated. Here, $\overline{x_{ij}}$ is the normalized matrix values and w is the criterion weight.

$$\widehat{x_{ij}} = \overline{x_{ij}} \times w \tag{3.6}$$

According to the qualitative nature of the output responses, maximizing index (P_j) and minimizing index (R_j) are calculated by Eqs. 3.7 and 3.8. Here, $\widehat{x_{ij}}$ is the weighted normalized matrix.

$$P_j = \sum \widehat{x_{ij}} \tag{3.7}$$

$$R_j = \sum \widehat{x_{ij}} \tag{3.8}$$

According to Eq. 3.9, the relative weights of the output responses are calculated.

$$Q_j = P_j + \frac{\sum R_j}{R_j \sum \frac{1}{R_j}} \tag{3.9}$$

According to Eq. 3.10, the degree of benefit is calculated. Here, major weight is Q_i: the longer is rank of work and Q_{max}: agreement degree is the largest.

$$N = \frac{Q_i}{Q_{max}} \times 100\% \tag{3.10}$$

Some studies on machinability related to COPRAS have been carried out. Nimel et al. optimized different cutting fluids (flood, MQL, CO_2 and CO_2+MQL), cutting speeds (60, 75 and 90 m/min) and feeds (0.04, 0.06 and 0.08 mm/rev) independent variables using the COPRAS technique in milling of Nimonic-80A material. They calculated machining temperature (MT), SR and flank wear (FW) dependent variables. It was found that the hybrid CO_2+MQL cooling technique was more effective than other methods to improve the C/L conditions during the processing of Nimonic-80A. It was revealed that flood cooling had better performance when compared to MQL due to the low penetrability of mist at high cutting speeds. It was determined that CO_2 cooling reduced the temperature of the cutting zone to the maximum level. For this reason, it was stated that better machined surface quality was obtained and FW was reduced. While it was found that successful results were obtained with the COPRAS technique, the optimum cutting conditions were calculated as hybrid CO_2+MQL, low cutting speed and feed [20]. Varatharajulu et al. optimized different spindle speeds (1100, 2920 and 4540 rpm) and feed rates (0.038, 0.076 and 0.203 mm/rev) using COPRAS and TOPSIS techniques in drilling of AZ91 magnesium. They measured drilling time, burr height and burr thickness as experimental results. The optimum parameters for simultaneous minimization of all experimental outputs using COPRAS and TOPSIS were calculated as spindle speed of 4540 rpm and feed rate of 0.076 mm/rev. Drilling alternatives were ranked in both statistical methods, and the results were evaluated. As a result, identical sequencing order was observed (Figure 3.5) [21].

Nguyen and Trung optimized the workpiece speed and depth of cut for minimum SR and maximum MRR in surface grinding of SKD11 steel using Taguchi, COPRAS and MOORA techniques. The optimum values for the independent variables were calculated as workpiece speed of 20 m/min and depth of cut of 0.02 mm. The experimental output values of SR and MRR for optimum points were 1.16 µm and 86.67 mm³/s, respectively. In addition, similar optimum cutting

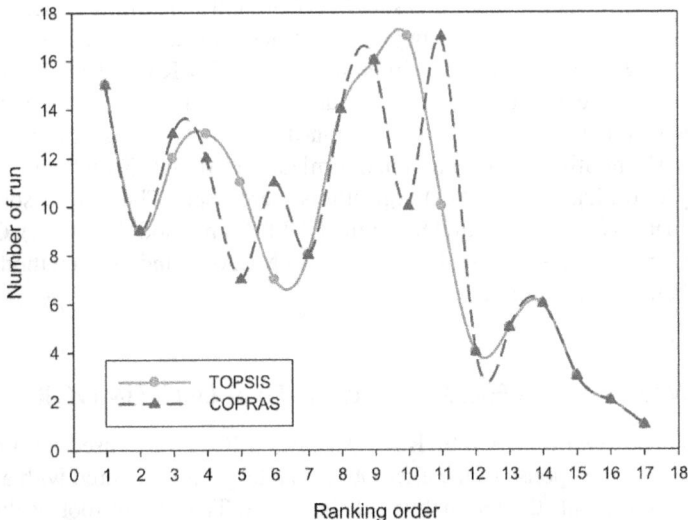

FIGURE 3.5 Variation between TOPSIS and COPRAS methods in terms of number of run/ranking order [21].

TABLE 3.1

Deviation in Different Metaheuristics Solutions for the Best Solution of COPRAS [25]

Metaheuristic Method	COPRAS Solution		% Deviation with Respect to Best Solution		Average Deviation
	MRR	Ra	MRR	Ra	
NSGA-II	0.00998	4.08457	0.62%	0.00%	0.31%
MOALO	0.01004	4.08499	0.00%	0.01%	0.01%
MODA	0.00991	4.08558	1.28%	0.02%	0.65%

parameters were determined with the statistical techniques used in the study [22]. Saha and Majumder found optimal points for different cutting speeds (160, 240 and 400 rpm), feeds (0.08, 0.16 and 0.32 mm/rev) and depths of cut (0.1, 0.15 and 0.2 mm) in turning of ASTM A36 steel using the COPRAS method. They calculated SR, power consumption and frequency of tool vibration as experimental output. Optimum parameters for minimum experimental output values were determined as spindle speed: 160 rpm, feed rate: 0.08 mm/rev and depth of cut: 0.1 mm. It was stated that the proposed statistical method could be used to solve it relatively easy compared to other traditional methods because it required less computation [23]. Stanojkovic and Radovanovic selected solid carbide drills for high pressured drilling with entropy and COPRAS method. While the alternatives were selected as Iscar, Seco, Sandvik and Kennametal, the selection criteria were determined as cutting speed, feed, pressured coolant and machining time. According to alternatives and criteria, ranking of preference was determined as Seco, Iscar, Sandvik and Kennametal [24]. Joshi et al. optimized MRR and SR in the parameters of cutting speed, feed and depth of cut in micro-turning of C360 copper alloy using tungsten carbide insert. For this, non-dominated sorting genetic algorithm II (NSGA-II), multi-objective ant lion optimization (MOALO) and multi-objective dragonfly optimization (MODA) algorithms were used. The Pareto solutions of these algorithms were compared using the COPRAS method (Table 3.1). COPRAS solutions for MODA were found to outperform MOALO and both methods outperformed from NSGA-II [25].

3.3.3 MOORA Plus Full Multiplicative Form (MULTIMOORA)

Multi-Objective Optimization by Ratio Analysis (MOORA) is used for a ratio system in which the response of the alternative on a target is compared with a denominator representing all the alternatives of the target. The square root of the sum of the squares of each alternative is chosen per objective for this denominator. This method [26] presented by Brauers is preferred for selecting alternatives. It works according to the ratio and reference point system. Desired and undesirable criteria in the ranking process are used simultaneously. It is preferred in terms of calculation system and ease. The full multiplicative form was added to MOORA by Brauers and Zavadskas and MULTIMOORA was created. Thus, MULTIMOORA consists of ratio system, reference point techniques and full multiplicative form [27]. It is widely used in solving problems encountered in different fields, such as MOORA [8]. In this method, after the standard decision matrix is created, the matrix is normalized (Eq. 3.11). Here, x_i is the criterion value.

$$X^* = \frac{x_i}{\sqrt{\sum x_i^2}}$$
(3.11)

According to Eq. 3.12, the normalized weight matrix is calculated. Here, x_i is the normalized matrix value and w is the criterion weight.

$$Y = x_i \times w$$
(3.12)

According to Eq. 3.13, the difference between the maximum and minimum values for each alternative is determined. According to Eq.3.14, the alternative value is calculated.

$$y_i = \sum \max x_{ij}^* - \sum \min x_{ij}^*$$
(3.13)

$$U_i = \frac{\sum \max X_{ij}^*}{\sum \min X_{ij}^*}$$
(3.14)

It was determined that the process parameters are optimized in the machining of materials using MOORA. Majumder and Maity optimized the microhardness and SR of shape memory alloy nitinol in the Wire electrical discharge machining (WEDM) process using general regression neural network (GRNN) and MOORA-fuzzy techniques. Different T_{on} (10, 12 and 14 µs), discharge current-I (8, 10 and 12 A), wire tension-WT (10, 12 and 14 N), wire speed-WS (150, 195 and 240 mm/s) and Flushing pressure (FP) (6, 8 and 10 bar) as independent variable were used. It was stated that the experimental outputs could be estimated with a ±10% error rate with the GRNN model. According to the MOORA-fuzzy MCDM approach, the optimum independent variables were determined as T_{on}: 12 µs, I: 10 A, WT: 12 N, WS: 150 mm/s and FP: 8 bar (Table 3.2). It was determined that the I was the most important variable on the experimental outputs with ANOVA results [28].

Anitha and Das optimized the machining parameters in the EDM process with the MOORA method. The effects of current, T_{on}, duty cycle and voltage-independent variables on MRR and SR were revealed. The improvement procedure was carried out using the combination of standard deviation and MOORA. Weights were calculated using the standard deviation of 0.53 for the MRR and 0.47 for the Ra. Optimal values to maximize MRR and minimize SR were calculated as current: 15 units, duty cycle: 50 units, T_{on}: 100 units and voltage 50 units [29]. Khan and Maity focused on the MOORA technique in WEDM, plasma arc cutting, electrochemical micro-machining, electrochemical machining, abrasive

TABLE 3.2
Confirmation Test Results [28]

Parameter level	Initial Machining Parameter $T_{ONI}\,I_I WS_I\,WT_I\,FP_{II}$	Optimum Machining Parameter Predicted $T_{ONII}\,I_{II}\,WT_{II}WS_I\,FP_{II}$	Experimental $T_{ONII}\,I_{II}\,WT_{II}WS_I\,FP_{II}$
Ra	3.78	2.31	2.27
Rq	4.64	4.33	4.12
Rz	22.4	20.18	19.77
MH	413.40	424.28	427.15
MFRG	0.686	0.89792	0.75
Improvement in MFRG	9.33%		

jet machining, abrasive water jet machining, ultrasonic machining, laser beam machining and laser cutting processes. They stated that this method could be used for large selection problems involving any amount of selection criteria in terms of relatively accurate, time saving and ease of operation [30]. Abhang et al. optimized the machining parameters for turning of EN-31 steel alloy with the MOORA technique. While different cutting speeds (39, 112 and 189 m/min), feeds (0.06, 0.1 and 0.15 mm/rev), cutting tool nose radii (0.2, 0.4 and 0.6 mm) and depths of cut (0.4, 0.8 and 1.2 mm) were used as independent variables, temperature, cutting force and tool wear were used as the dependent variables or experimental outputs. It was stated that all independent variables used for the dependent in the machining of EN-31 steel alloy should be used at the lowest values. Gray relational analysis and MOORA results were found to be compatible with each other [31]. Sahu et al. optimized the SR of AISI 1040 stainless steel with MOORA in the EDM process, using selective laser sintering (SLS) manufactured AlSiMg, standard copper and brass electrodes. T_{on}: 100, 200 and 300 µs and discharge current (Ip: 10, 20 and 30 A) were used as independent variables. According to the MOORA optimization, it was revealed that better surface quality was obtained by using AlSiMg electrode, Ip: 10 A and T_{on}: 100 µs parameters under the same conditions. In other words, it was determined that the lowest Ip and T_{on} values and the electrode manufactured with SLS should be used in terms of better surface quality [32]. Sahoo et al. optimized the experimental parameters in the WEDM process of High carbon high chromium steel with the MOORA technique. MRR, KW and average SR experimental outputs were obtained using different pulse width time (T_{on}: 15, 20 and 25 µs), pulse off time (T_{off}: 25, 30 and 35 µs) and wire feed rate (WF: 8, 10 and 12 m/min) parameters. It was determined that MRR, KW and SR increased with increasing of T_{on} and decreased more slowly with increasing of T_{off}. It was revealed that T_{on}: 15 µs, T_{off}: 35 µs and WF: 10 m/min should be used to maximize MRR and minimize KW and SR with MOORA. It was also stated that these results were the best combination for efficient machining work in industries [33].

3.3.4 WEIGHTED AGGREGATED SUM PRODUCT ASSESSMENT (WASPAS)

WASPAS is a multi-response appropriate decision-making method. A mutual optimality criterion is determined based on two optimality criteria in this method. It is used to evaluate a set of alternatives according to a set of decision criteria. It is quite practical and utilizes heavily on the concept of ranking accuracy. It is preferred in multiple response systems in terms of different engineering fields to find the optimum parametric setting for combined output responses [34]. The WASPAS technique is a unique combination of two commonly used MCDM techniques. In other words, WASPAS, which combines the Weighted Sum Model (WSM) and the Weighted Product Model (WPM), increases the ranking accuracy of the alternatives [35, 36]. $\left[x_{ij} \right]_{n \times m}$ decision matrix consisted of according to Eq. 3.15 with "n" alternative and "m" criteria in the first stage.

$$\left[x_{ij} \right]_{n \times m} = \begin{bmatrix} x_{11} & x_{12} & \cdots & x_{1m} \\ x_{21} & x_{22} & \cdots & x_{2m} \\ \cdots & \cdots & \cdots & \cdots \\ x_{n1} & x_{n2} & \cdots & x_{nm} \end{bmatrix} \tag{3.15}$$

According to Eq. 3.16, the decision matrix is normalized in the second stage. Here, $\bar{X}_{ij}$ is normalized decision matrix. N_b and N_{nb} represent benefit and non-benefit criteria, respectively. N_b is desired to be large, while N_{nb} is desired to be small.

$$\left(\bar{X}_{ij} \right) = \begin{cases} \dfrac{x_{ij}}{\max x_{ij}} & j \in N_b \\ \dfrac{\min x_{ij}}{x_{ij}} & j \in N_{nb} \end{cases} \tag{3.16}$$

According to Eq. 3.17, the relative importance of the alternatives is calculated using the WSM method. Here, W_j: is the weight value of the j criterion. According to Eq. 3.18, the total relative importance of the i. alternative using the WPM method is calculated.

$$Q_i^{(1)} = \sum_{j=1}^{n} \bar{X}_{ij} W_j \tag{3.17}$$

$$Q_i^{(2)} = \prod_{j=1}^{n} \left(\bar{X}_{ij} \right)^{w_j} \tag{3.18}$$

As a result of the weighted combination of additive and multiplicative methods, the mutual generalized criterion is calculated as in Eq. 3.19 [37].

$$Q_i = 0.5 Q_i^{(1)} + 0.5 Q_i^{(2)} \tag{3.19}$$

Total relative importance of the i. alternative for improved and reliable ranking accuracy in WASPAS is determined by Eq. 3.20 [38].

$$Q_i = \lambda Q_i^{(1)} + (1 - \lambda) Q_i^{(2)} \tag{3.20}$$

All alternatives are ranked according to their Q values. The highest Q value has the best alternative. The WASPAS method turns into WPM in the case of $\lambda = 0$ and WSM in the case of $\lambda = 1$ [39–42]. It has been determined that there are some studies in the literature about WASPAS. Pathapalli et al. optimized the machining parameters in turning of stir casted and TiC reinforced Al6060 metal matrix composite with the WASPAS technique. Optimum parameters for maximum MRR, minimum cutting force and SR were determined using different input variables of speed (700, 950 and 1200 rpm), feed rates (0.1, 0.13 and 0.16 mm/rev), depths of cut (0.1, 0.2 and 0.3 mm) and TiC reinforcement rates (5, 10 and 15 Wt.%). It was determined that all experimental outputs were affected by speed, feed rate and reinforcement ratio. While MRR and SR were affected from speed by 53.41% and 47.75%, respectively. Cutting force was affected from feed rate by 59.25%. It was observed that WASPAS and MOORA exhibited similar results [39]. Reddy et al. optimized the AWJM process parameters for the Inconel-625 alloy using WASPAS and MOORA. While different SOD (1, 1.5 and 2 mm), traverse speed (97, 117 and 146 mm/s) and sand flow rate (200, 220 and 250 g/min) were used for independent variables, MMR, KW and SR were used for dependent variables. It was determined that MRR increased with the increase of traverse speed and abrasive flow rate, and while SR increased with the increase of abrasive flow rate, it decreased with the increase of traverse speed. While KW increased with increasing of SOD, it decreased with increasing of traverse speed. It was found that the ranking of the alternatives obtained with WASPAS and MOORA was the same [40]. Prasad et al. optimized AJM process parameters of nickel 233 alloy with WASPAS and MOORA using different pressure (5, 6 and 7 kgf/cm^2), abrasive grain size (300, 400 and 500 μm) and SOD (5, 7 and 9 mm) according to MRR, SR and T_a outputs. Optimum output values with WASPAS were determined as maximum MRR: 1 mg/min, minimum SR: 0.5679 μm and T_a: 0.2977°. According to the results, it was observed that there was a parallelism between WASPAS and MOORA [43]. Tudu et al. optimized the independent variable parameters in the EDM and WEDM processes of Ti-6Al-4V alloy with WASPAS and multi-objective genetic algorithm (MOGA). Experimental outputs were calculated as MRR and SR, while the independent variables were used as T_{on}, T_{off}, W_f and wire tension (W_t). It was revealed that MOGA was more flexible for constrained and unconstrained complex integer problems to find optimum parameters [44]. Sahoo et al. optimized spindle speeds (600, 650 and 700 rpm), feeds (0.25, 0.375 and 0.5 mm/min) and depths of cut (0.2, 0.3 and 0.4 mm) variables with WASPAS according to tool vibration (V) and SR in turning of Al6063-T6 alloy. Optimum cutting parameters were calculated as spindle speed: 600 rpm, feed rate: 25 mm/min and depth of cut: 0.3 mm for minimum V: 69.580 dB and SR: 0.596 μm [45].

3.3.5 EVALUATION OF MIXED DATA (EVAMIX)

EVAMIX, which is among the decision-making analysis approach systems, reduces the selection time or decision-making process by using complex, independent quantitative and qualitative features. Because of these features, EVAMIX is flexible and different from other methods [46]. It is among the compensatory methods and the qualifications are independent of each other. It is not necessary to convert the qualitative features into quantitative features in this method. The input information of EVAMIX is explained using a matrix of alternatives and attributes based on information obtained from the decision-maker [47]. Firstly, the $(m \times n)$ size decision matrix is created in this technique and the weights of the criteria are calculated [48]. Here, n is the number of alternatives and m is the number of relative features selected [49]. In the next step, linear normalization is performed in the range of 0–1. The dominance scores of each pair of alternatives (i, i') are calculated for all rank and main criteria using Eqs. 3.21–3.23. Here, the features are divided into two categories as ordinal (O) and cardinal (C). Here r_{ij} refers to the evaluation of alternative A_i based on attribute C_j. $r_{i'j}$ is the evaluation of alternative A_i' based on the attribute C_j. W_j is the weight attributed to the feature calculated by both techniques. Cardinal attributes are compared with the calculation of the dominant factor $a_{ii'}$, which indicates the superiority of A_i to A_i'. $a_{ii'}$ is calculated according to Eq. 3.21.

$$a_{ii'} = \left[\sum_{j \in O} \left\{ W_j \operatorname{sgn}(r_{ij} - r_{i'j}) \right\}^c \right]^{\frac{1}{c}} \tag{3.21}$$

$$\operatorname{Sgn}(r_{ij} - r_{i'j}) = \begin{cases} +1 & r_{ij} > r_{i'j} \\ 0 & r_{ij} = r_{i'j} \\ -1 & r_{ij} < r_{i'j} \end{cases} \tag{3.22}$$

$$\gamma_{ii'} = \left[\sum_{j \in C} \left\{ W_j \operatorname{sgn}(r_{ij} - r_{i'j}) \right\}^c \right]^{\frac{1}{c}} \tag{3.23}$$

$a_{ii'}$ and $\gamma_{ii'}$ are dominance scores for alternative pairwise (ii') scores according to ordinal and cardinal criteria. Since $a_{ii'}$ and $\gamma_{ii'}$ will have different units of measure, a standardization in the same unit is required. Standardized dominance scores can be written as Eq. 3.24.

$$\delta_{ii'} = h(a_{ii'}) \text{ve } d_{ii'} = h(\gamma_{ii'}) \tag{3.24}$$

h represents a standardization function. Standardized dominance scores can be obtained using three different approaches. Standardized rank score $(\delta_{ii'})$ and cardinal dominance score $(d_{ii'})$ for the alternative pair (i, i') using the additive interval technique is calculated according to Eqs. 3.25 and 3.26.

$$(\delta_{ii'}) = \left(\frac{a_{ii'} - a^-}{a^+ - a^-} \right) \tag{3.25}$$

Standardized cardinal dominance score in case of the highest rank dominance score for the alternative pair is calculated according to Eq. 3.26. Here, $\gamma^+ (\gamma^-)$ is the highest (lowest) cardinal dominance score for the alternative pair (i, i').

$$(d_{ii'}) = \left(\frac{\gamma_{ii'} - \gamma^-}{\gamma^+ - \gamma^-} \right) \tag{3.26}$$

Assuming that the w_j weights have quantitative properties, the general dominance scale $D_{ii'}$ for each pair of (i, i') alternatives is calculated according to Eq. 3.27.

$$D_{ii'} = w_O \delta_{ii'} + w_C d_{ii'} \qquad (3.27)$$

w_O is the sum of the weights for ordinal criteria ($w_O = \Sigma_{j \in O} w_j$). w_C is the sum of the weights for the cardinal criteria ($w_C = \Sigma_{j \in C} w_j$). This overall dominance score indicates the degree of dominance of a_i over $a_{i'}$ for the given feature set and weights. In other words, the $D_{ii'}$ can be accepted as a k function of the founder evaluation scores according to Eq. 3.28.

$$D_{ii'} = k\left(s_i, s_{i'}\right) \qquad (3.28)$$

In the final stage, the evaluation score is calculated for the alternative (S_i). According to this result, the final preference of the candidate alternatives is determined. The best alternative is the one with the highest value of the evaluation score. Accordingly, the evaluation score of the alternatives is calculated according to Eq. 3.29. The resulting S_i values are ranked in descending order [48–50].

$$\left(S_i\right) = \sum_{i'} \left(\frac{D_{i'i}}{D_{ii'}}\right)^{-1} \qquad (3.29)$$

It was determined that the material selection has been made for the appropriate manufacturing conditions using EVAMIX in the literature [48, 51]. In another study, Chatterjee and Chakraborty found optimum non-traditional machining (NTM) technique using EVAMIX. Criteria affecting the NTM selection decision were determined as tolerance and surface quality (TSF), power requirement (PR), MRR, cost (C), tooling and fixtures (TF), tool consumption (TC)), safety (S), workpiece material (M) and shape property (F). In the first example, the Electrochemical machining (ECM) method was in the first place in the machining of a stainless steel material according to the ranking, while Abrasive jet machining (AJM) was in the last place. The best result was obtained with Electrochemical machining (EDM in the machining of aluminum alloy, while the worst result was obtained with AJM. The best result was obtained with USM for ceramic materials, while the worst result was obtained with the ECM process. They also stated that the EVAMIX algorithm provided much more flexibility than other MCDM methods due to the characteristics of cardinal and ordinal data designed to combine the output into a single evaluation score[52].

3.3.6   OCCUPATIONAL REPETITIVE ACTIONS (OCRA)

OCRA was developed by Parkan in 1991 to calculate the performance of alternatives in performance and efficiency measurement and analysis problems. It uses a heuristic approach to combine the decision-maker's preferences with the relative importance of the criteria [53]. The main purpose of the OCRA method is to evaluate the alternatives based on benefit and non-benefit criteria and to obtain the final ranking by combining these two classification sets [54]. The relative weights of the criteria depend on the alternatives in the OCRA method. Different alternatives in different weight distributions are assigned to the criteria. It can be applied to all alternatives. Benefit (maximization) and non-benefit (minimization) criteria can be handled separately in this method. This helps decision-makers not to lose information in the decision-making process. Another important advantage of the OCRA method is that it is a non-parametric approach. In other words, the calculation procedure is not affected by the addition of any additional parameters, as in other MCDM methods. The result can be achieved with less calculation processes. While only six steps are required to solve a particular decision-making problem using the OCRA method, nine steps are required in the TOPSIS method [54]. According to Eq. 3.30, decision matrix is formed in the OCRA method as in other methods. Here, a_{ij}, represents the performance value of i alternative in the j criterion. m and n are the number of alternatives and criteria, respectively.

$$A = \left[a_{ij}\right]_{m\times n} = \begin{bmatrix} a_{11} & a_{12} & \cdots & a_{1n} \\ a_{21} & a_{22} & \cdots & a_{2n} \\ \cdots & \cdots & \cdots & \cdots \\ a_{m1} & a_{m2} & \cdots & a_{mn} \end{bmatrix} \tag{3.30}$$

Preference ratings are determined according to non-benefit criteria. According to all non-benefit criteria, the total performance of the i. alternative is calculated. According to criteria, the ranking of preference is determined by Eq. 3.31. Here, $\bar{I}_i$ is the i. the relative performance value of the alternative according to the non-benefit criteria, q: number of non-benefit criteria, a_i^k: the performance score of i alternative in the k criterion and w_k: the weight of the non-benefit k criterion. When i alternative is preferred according to the k criterion unlike the m alternative, it is accepted as $a_i^k < a_m^k$.

$$\bar{I}_i = \sum_{k=1}^{q} w_k \frac{\max\left(a_i^k\right) - a_i^k}{\min\left(a_i^k\right)}, (i = 1,2,\ldots, m) \tag{3.31}$$

According to Eq. 3.32, linear preference ranking is performed for non-benefit criteria. Here, according to non-benefit criteria, the value of $\bar{\bar{I}}_i$ is the total ranking of preferences for the i. alternative.

$$\bar{\bar{I}}_i = \bar{I}_i - \min\left(\bar{I}_i\right) \tag{3.32}$$

According to benefit criteria, preference rankings are calculated using Eq. 3.33. Here, $\bar{O}_i$: the relative performance value of the i alternative with respect to useful criteria, b: the number of benefit criteria, and w_h: the weight of the benefit h criterion. The higher an alternative score for a useful criterion, the higher the preference for that alternative.

$$\bar{O}_i = \sum_{h=1}^{b} w_h \frac{a_i^h - \min\left(a_i^h\right)}{\min\left(a_i^h\right)}, (i = 1,2,\ldots, m) \tag{3.33}$$

Eq. 3.34 is used for determining linear preference ranking for benefit criteria. Here $\bar{O}_i$ value, refers to the total preference ranking for the i. alternative according to benefit criteria.

$$\bar{\bar{O}}_i = \bar{O}_i - \min\left(\bar{O}_i\right) \tag{3.34}$$

According to Eq. 3.35, the general preference ranking is calculated at the last stage. Here, P_i represents the total preference value and this value is obtained for each alternative. Thus, the full ranking of the alternatives is revealed based on the general preference stages. The alternative with the highest general performance rating is in first rank. The alternative with the highest total preference value is determined as the best alternative [54, 55].

$$P_i = \left(\bar{\bar{I}}_i + \bar{\bar{O}}_i\right) - \min\left(\bar{\bar{I}}_m + \bar{\bar{O}}_m\right) \tag{3.35}$$

In studies on the OCRA method, Madic et al. used the OCRA method for optimum non-conventional machining technique selection [54]. Non-conventional machining technique selection was also made using different MCDM methods in [56] and [57]. As a result, it was stated that a good correlation was obtained with the study results of the mentioned authors [54]. A similar finding was also found in [58]. Patel and Maniya optimized the process parameters in the WEDM process of Al-SiC composite material with OCRA. They used different T_{on} (108, 115, 123 and 130 µs), T_{off} (50, 54, 58 and 62 µs), wire diameter (0.25 and 0.3 mm), uncoated and zinc coated wire as independent variables. MRR, SR and cutting velocity (CV) were calculated as experimental results. Optimum parameters for uncoated electrode wire were measured as T_{on}: 123 µs, T_{off}: 58 µs and wire diameter: 0.25 mm. Optimum parameters for zinc-coated electrode wire were calculated as T_{on}: 130 µs, T_{off}: 50 µs and wire diameter: 0.3 mm. While MRR and CV improved under the same conditions with zinc-coated wire, SR increased with uncoated wire. Small wire diameter increased MRR and CV, while large wire diameter increased machined SR [59]. Kumar and Ray selected the optimum gear material with OCRA, EXPROM-2, ORESTE and MOOSRA. Alternatives were selected as cast iron, ductile iron, Spheroidal graphite (S.G.) iron, cast alloy steel, through hardened alloy steel, surface hardened alloy steel, carburized steels, nitrided steels and through hardened carbon steel. Criteria were determined as surface hardness, core hardness, surface fatigue limit, bending fatigue limit and ultimate tensile strength. It was determined that OCRA and EXPROM-2 techniques gave better results than other MCDM methods and the calculation process was easier [60]. Patel and Maniya optimized the WEDM process parameters of Al6061 metal matrix composite material different reinforced (SiC, B4C and ZrO_2) using OCRA, ARAS (Additive ratio assessment), TOPSIS, MOORA and GRA techniques. For this, they used different wire materials (copper, brass and molybdenum), wire diameters (0.2, 0.25 and 0.3 mm), T_{on} (108, 118 and 128 µs), T_{off} (50, 55 and 60 µs), PCs (100, 150 and 200 amp) wire tensions (4, 7 and 10 kgf) and wire feeds (3, 6 and 9 m/min). MRR, CV and SR were measured as experimental output. Optimal parameters in SiC reinforced material for maximum MRR and CV, and minimum SR were determined brass wire material, wire diameter of 0.25 mm, T_{on} of 128 μs, T_{off} of 50 μs, PC of 150 amp, wire tension of 10 kgf, wire feed rate of 3 mm/min. The optimum parameters for B4C and ZrO_2 reinforced material were determined as brass wire material, wire diameter of 0.25 mm, T_{on} of 128 μs, T_{off} of 50 μs, PC of 100 amp, wire tension of 7 kgf, wire feed rate of 9 mm/min. In addition, it was found that the proposed ranking methods were successful to reveal the best alternative [61].

3.3.7 MULTI-ATTRIBUTIVE BORDER APPROXIMATION AREA COMPARISON (MABAC)

It is the new MCDM method used by Pamucar and Cirovicfoth for rational decision-making and obtaining consistent solutions. This method is capable of calculating the distance measure between each alternative and the border approximation area and making reasonable decisions. It is recommended as one folded because it provides stable solutions compared to other MCDM methods. In addition, this method takes into account hidden gain and loss values and can be used as a hybrid with other approaches [62–64]. In the first stage, a decision matrix is created according to Eq. 3.36, and the weight of the criteria is decided for this method. Here, m and n are the number of alternatives and criteria, respectively.

$$A = \begin{bmatrix} a_{11} & a_{12} & \cdots & a_{1n} \\ a_{21} & a_{22} & \cdots & a_{2n} \\ \cdots & \cdots & \cdots & \cdots \\ a_{m1} & a_{m2} & \cdots & a_{mn} \end{bmatrix} \tag{3.36}$$

In the second step, the decision matrix is normalized for easy comparison and conversion to dimensionless. For each criterion, the maximum value $a_{j,max}$ and the minimum value $a_{j,min}$ are calculated. The decision matrix is normalized according to Eq. 3.37 and Eq. 3.38 for the benefit and cost type criteria, respectively. Here, a_{ij} is performance measurement of the i alternative of j criterion.

$$a_{ij}^n = \frac{a_{ij} - a_{j,min}}{a_{j,max} - a_{j,min}} \tag{3.37}$$

$$a_{ij}^n = \frac{a_{j,max} - a_{ij}}{a_{j,max} - a_{j,min}} \tag{3.38}$$

The weighted normalized decision matrix is created by calculating according to $a_{ij}^w = w_j + a_{ij}^n w_j$ mathematical formula. w_j is the weight of j criterion in this formula. Border approximation area matrix values are calculated according to Eq. 3.39.

$$BAA_j = \left(\prod_{i=1}^{m} a_{ij}^w \right)^{\frac{1}{m}} \quad j:1,\ldots\ldots,n \tag{3.39}$$

Distance from border approximation of alternatives area is determined according to
Eq. 3.40. The total distance of each alternative from the border approximation area
is calculated according to Eq. 3.41. In the last step, the total distances of the alternatives from the border approximation area are determined in descending rank and the
final ranking of the alternatives is created [64–67].

$$q_{ij} = a_{ij}^{w} - BAA_J \quad i:1, \ldots\ldots, m, \ j:1,\ldots\ldots,n \tag{3.40}$$

$$S_i = \sum_{j=1}^{n} q_{ij}; i:1, \ldots\ldots, m \tag{3.41}$$

In studies about MABAC, Paramasivam et al. optimized with MABAC the machining parameters of AM-60 magnesium alloy under different conditions using a cryo-
genic-treated tool. Different cutting speeds (2000, 2500 and 3000 rpm), feeds (0.24,
0.3 and 0.36 mm/rev) and drill bit treatment were used as input parameters, while
output parameters were MMR, SR, PR, feed force, burr height and circularity error
(Figure 3.6). It was stated that the result classification scheme was used to optimize
the machining parameters with MABAC. Therefore, it was revealed that it was a
simple and useful tool for practical decision-making problems. The optimum parameters with this method were calculated as a cutting speed of 3000 rpm, feed rate of
0.36 mm/rev and two tempering cycled tools [68].

Shivakoti et al. optimized T_{on}, T_{off}, wire tension and servo voltage according to CS
and SR in the WEDM process of D3 die steel using MABAC. Optimum parameters
for minimum SR and maximum cutting speed were determined as servo voltage: 40 V,
T_{on}: 115 μs, T_{off}: 45 μs and wire tension: 7 kgf. According to the optimum points, the
SR and cutting speed values were calculated as 1.556 μm and 2.54 mm/min, respectively [64]. Chatterjee et al. studied the selection of unconventional machining processes with FARE (Factor relationship)-MABAC. When the results were compared

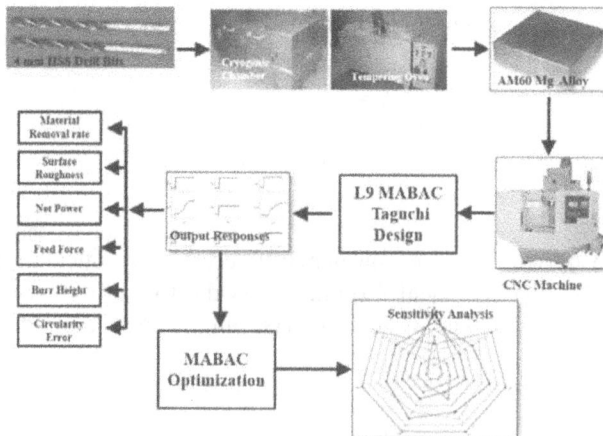

FIGURE 3.6   Block schema for MABAC [68].

with the reference [69], it was revealed that it was quite compatible with the results of the current algorithm. It was stated that it could contribute to the decision-making of engineers and designers in the selection of the best alternative [70].

3.4 CONCLUDING REMARKS

There are many different types of MCDM used in the literature. Different methods are preferred according to different sectoral applications. Many MCDM techniques continue to be developed according to the needs. VIKOR is a consensus-based ranking method that provides the closest decision to the ideal solution under certain conditions by creating a multi-criteria ranking index among the alternatives. COPRAS compares the alternatives with each other and reveals how good or how bad they are compared to the other alternatives as a percentage. As a result, it contributes to the ranking of alternatives. It shows the degree of utility of the alternatives and can be used with other methods for criterion weighting. The COPRAS method shows the degree of benefit of the alternatives and can be used with other methods for criterion weighting. Calculations can be performed for both maximizing and minimizing criteria. The criteria can be handled and evaluated separately during the evaluation process. MULTIMOORA is a method based on numerical data. It is a process of simultaneous optimization of alternatives under certain constraints. The results obtained by using the MOORA method and Fully Multiplicative Form are combined according to their order-dominance status. It is more robust than other methods based on ordinal measures. It is possible to work on positive data. WASPAS is preferred in applications where high consistency is desired in the ranking of alternatives by increasing the ranking accuracies and optimizing the calculated weighted integrated function. The ranking of alternatives according to performance is taken into account in the OCRA method. Criteria including numerical and verbal data with EVAMIX can be examined at the same time and used to rank the alternatives. The MABAC method is used to determine the most suitable alternatives. It handles complex and uncertain decision-making problems by calculating the distance between each alternative and the boundary approach area. MCDM comes to the fore in subjects and engineering applications where many options have an impact on the process. The concepts of cost, quality and time come to the fore in the manufacturing industry. In this context, manufacturing methods include many variables. For this reason, MCDM is preferred as an assist tool in determining the most appropriate performance conditions by reducing cost and time in manufacturing. In particular, different results can be obtained depending on the cutting parameters inputs in machining techniques. In this context, it is necessary to determine the suitability of processing parameters such as T_{on}, T_{off}, SOD, PC, wire diameter, wire type, servo voltage, pressure, abrasive grain size, feed, cutting speed and depth of cut which can affect the result in cutting processes. For this, it is possible to avoid unnecessary cutting operations by first analyzing experimental outputs such as MRR, SR, KW, cutting force and T_a with MCDM methods. The aim of the analysis is to obtain the best solution that provides the manufacturing conditions. On the other hand, MCDM accelerates decision-making processes and increases decision quality. Effective, accurate and real-time solutions of real-life problems can be realized by using these methods. Results with high accuracy values are obtained with mathematical models that best reflect the characteristics of the cutting process and their interactions with other factors in the environment using MCDM methods. Thus, it is possible to prefer MCDM techniques in different disciplines.

REFERENCES

1. Eltarabishi, F., Omar, O.H., Alsyouf, I., & Bettayeb, M. (2020). Multi-Criteria decision making methods and their applications– A literature review. *Proceedings of the International Conference on Industrial Engineering and Operations Management*, Dubai, UAE (pp. 2654–2663).
2. Bayraktar, S., & Gupta, K. (2021). *Multi-criteria decision making through soft computing and evolutionary techniques*. In *Intelligent Manufacturing* (pp. 123–147). Cham: Springer.
3. Castro, D.M., & Parreiras, F.S. (2021). A review on multi-criteria decision-making for energy efficiency in automotive engineering. *Applied Computing and Informatics*, 17(1), 53–78.
4. Melnik-Leroy, G.A., & Dzemyda, G. (2021). How to influence the results of MCDM? – Evidence of the impact of cognitive biases. *Mathematics*, 9(2), 121.
5. Mostafaeipour, A., Dehshiri, S.S.H., Dehshiri, S.J.H., Almutairi, K., Taher, R., Issakhov, A., & Techato, K. (2021). A thorough analysis of renewable hydrogen projects development in Uzbekistan using MCDM methods. *International Journal of Hydrogen Energy*, 46(61), 31174–31190.
6. Bertolini, M., Esposito, G., & Romagnoli, G. (2020). A TOPSIS-based approach for the best match between manufacturing technologies and product specifications. *Expert Systems with Applications*, 159, 113610.
7. Chatterjee, P., & Chakraborty, S. (2016). A comparative analysis of VIKOR method and its variants. *Decision Science Letters*, 5(4), 469–486.
8. Siksnelyte-Butkiene, I., Streimikiene, D., Balezentis, T., & Skulskis, V. (2021). A systematic literature review of multi-criteria decision-making methods for sustainable selection of insulation materials in buildings. *Sustainability*, 13(2), 737.
9. Zlaugotne, B., Zihare, L., Balode, L., Kalnbalkite, A., Khabdullin, A., & Blumberga, D. (2020). Multi-criteria decision analysis methods comparison. *Scientific Journal of Riga Technical University. Environmental and Climate Technologies*, 24(1), 454–471.
10. Samson, R.M., Rajak, S., Kannan, T.D.B., & Sampreet, K. R. (2020). Optimization of process parameters in abrasive water jet machining of Inconel 718 using VIKOR method. *Journal of The Institution of Engineers (India): Series C*, 101(3), 579–585.
11. Singaravel, B., Prasad, S.D., Shekar, K.C., Rao, K.M., & Reddy, G.G. (2020). *Optimization of process parameters using hybrid Taguchi and VIKOR method in electrical discharge machining process*. In Advanced Engineering Optimization Through Intelligent Techniques (pp. 527–536). Singapore: Springer.
12. Dey, A., Shrivastav, M., & Kumar, P. (2021). Optimum performance evaluation during machining of Al6061/cenosphere AMCs using TOPSIS and VIKOR based multi-criteria approach. Proceedings of the Institution of Mechanical Engineers, Part B: Journal of Engineering Manufacture, 0954405420958770, 235(13), 2174–2188.
13. Khan, A., & Maity, K. (2020). Estimation of optimal cutting conditions during machining of CP-Ti grade 2 in fuzzy–VIKOR context: A hybrid approach. *Grey Systems: Theory and Application*, 10(3), 293–310.
14. Kumar, J., Soota, T., Rajput, S. K., & Saxena, K.K. (2021). Machining and optimization of Zircaloy-2 using different tool electrodes. *Materials and Manufacturing Processes*, 36(13), 1513–1523.

15. Vikram, K.A., Kanth, T.K., & Prasad, R.D.V. (2020). Response optimization using VIKOR while machining on lathe under dry and minimum quantity and lubrication conditions – A case study. *Materials Today: Proceedings*, 27, 2487–2491.

16. Huu Phan, N., & Muthuramalingam, T. (2021). Multi-criteria decision-making of vibration-aided machining for high silicon-carbon tool steel with Taguchi–TOPSIS approach. *Silicon*, 13(8), 2771–2783.

17. Babu, K.A., & Jeyapaul, R. (2021). Process parameters optimization of electrical discharge wire cutting on AA6082/Fly Ash/Al 2 O 3 hybrid MMC using Taguchi method coupled with hybrid approach. *Journal of The Institution of Engineers (India): Series C*, 102(1), 183–196.

18. Sahu, A.K., Thomas, J., & Mahapatra, S.S. (2021). An intelligent approach to optimize the electrical discharge machining of titanium alloy by simple optimization algorithm. *Proceedings of the Institution of Mechanical Engineers, Part E: Journal of Process Mechanical Engineering*, 235(2), 371–383.

19. Krishnaveni A., Jebakani D., Jeyakumar K., & Pitchipoo P. (2016). Turning parameters optimization using COPRAS-Taguchi technique. *International Journal of Advanced Engineering Technology*, 7, 463–478.

20. Mia, M., Anwar, S., Manimaran, G., Saleh, M., & Ahmad, S. (2021). A hybrid approach of cooling lubrication for sustainable and optimized machining of Ni-based industrial alloy. *Journal of Cleaner Production*, 321, 128987.

21. Varatharajulu, M., Duraiselvam, M., Kumar, M.B., Jayaprakash, G., & Baskar, N. (2021). Multi criteria decision making through TOPSIS and COPRAS on drilling parameters of magnesium AZ91. *Journal of Magnesium and Alloys*.

22. Nguyen, N.T., & Trung, D. (2021). Combination of Taguchi method, MOORA and COPRAS techniques in multi-objective optimization of surface grinding process. *Journal of Applied Engineering Science*, 19(2), 390–398.

23. Saha, A., & Majumder, H. (2016). Multi criteria selection of optimal machining parameter in turning operation using comprehensive grey complex proportional assessment method for ASTM A36. *International Journal of Engineering Research in Africa*, 23, 24–32.

24. Stanojkovic, J., & Radovanovic, M. (2017). Selection of drill for drilling with high pressure coolant using entropy and COPRAS MCDM method. *UPB Scientific Bulletin, Series D Mechanical Engineering*, 79, 199–204.

25. Joshi, M., Ghadai, R.K., Madhu, S., Kalita, K., & Gao, X.Z. (2021). Comparison of NSGA-II, MOALO and MODA for multi-objective optimization of micro-machining processes. *Materials*, 14(17), 5109.

26. Brauers, W.K. (2003). *Optimization methods for a stakeholder society: a revolution in economic thinking by multi-objective optimization* (Vol. 73). Norwell, Massachusetts, USA: Springer Science & Business Media.

27. Zavadskas, E.K., Antucheviciene, J., Razavi Hajiagha, S.H., & Hashemi, S.S. (2015). The interval-valued intuitionistic fuzzy MULTIMOORA method for group decision making in engineering. *Mathematical Problems in Engineering*, 2015, 1–13.

28. Majumder, H., & Maity, K. (2018). Prediction and optimization of surface roughness and micro-hardness using GRNN and MOORA-fuzzy-a MCDM approach for nitinol in WEDM. *Measurement*, 118, 1–13.

29. Anitha, J., & Das, R. (2021). *Optimization of process parameters in EDM using standard deviation and MOORA method*. In *Advances in Materials and Manufacturing Engineering* (pp. 151–158). Singapore: Springer.

30. Khan, A., & Maity, K.P. (2016). Parametric optimization of some non-conventional machining processes using MOORA method. *International Journal of Engineering Research in Africa*, 20, 19–40.

31. Abhang, L.B., Iqbal, M., & Hameedullah, M. (2020). Optimization of machining process parameters using MOORA method. *Defect and Diffusion Forum*, 402, 81–89.

32. Sahu, A.K., Mahapatra, S.S., Chatterjee, S., & Thomas, J. (2018). Optimization of surface roughness by MOORA method in EDM by electrode prepared via selective laser sintering process. *Materials Today: Proceedings*, 5(9), 19019–19026.

33. Sahoo, S.K., Naik, S.S., & Rana, J. (2019). *Experimental analysis of wire EDM process parameters for micromachining of high carbon high chromium steel by using MOORA technique.* In Micro and Nano Machining of Engineering Materials (pp. 137–148). Cham: Springer.

34. Pradhan, S., Indraneel, S., Sharma, R., Bagal, D.K., & Bathe, R.N. (2020). Optimization of machinability criteria during dry machining of Ti-2 with micro-groove cutting tool using WASPAS approach. *Materials Today: Proceedings*, 33, 5306–5312.

35. Senapati, T., Yager, R.R., & Chen, G. (2021). Cubic intuitionistic WASPAS technique and its application in multi-criteria decision-making. *Journal of Ambient Intelligence and Humanized Computing*, 12(9), 8823–8833.

36. Mathew, M., & Sahu, S. (2018). Comparison of new multi-criteria decision making methods for material handling equipment selection. *Management Science Letters*, 8(3), 139–150.

37. Šaparauskas, J., Kazimieras Zavadskas, E., & Turskis, Z. (2011). Selection of facade's alternatives of commercial and public buildings based on multiple criteria. *International Journal of Strategic Property Management*, 15(2), 189–203.

38. Zavadskas, E.K., Antuchevıcıene, J., Saparauskas, J., & Turskis, Z. (2013). MCDM methods WASPAS and MULTIMOORA: Verification of robustness of methods when assessing alternative solutions. *Economic Computation and Economic Cybernetics Studies and Research*, 47, 5–20.

39. Pathapalli, V.R., Basam, V.R., Gudimetta, S.K., & Koppula, M.R. (2020). Optimization of machining parameters using WASPAS and MOORA. *World Journal of Engineering*, 17(2), 237–246.

40. Reddy, P.V., Kumar, G.S., & Kumar, V.S. (2020). Multi-response optimization in machining Inconel-625 by abrasive water jet machining process using WASPAS and MOORA. *Arabian Journal for Science and Engineering*, 45(11), 9843–9857.

41. Baykasoğlu, A., & Gölcük, I. (2019). Revisiting ranking accuracy within WASPAS method. *Kybernetes*, 49(3), 885–895.

42. Zavadskas, E.K., Chakraborty, S., Bhattacharyya, O., & Antucheviciene, J. (2015). Application of WASPAS method as an optimization tool in non-traditional machining processes. *Information Technology and Control*, 44(1), 77–88.

43. Prasad, S.R., Ravindranath, K., & Devakumar, M.L.S. (2018). Experimental investigation and parametric optimization in abrasive jet machining on nickel 233 alloy using WASPAS and MOORA. *Cogent Engineering*, 5(1), 1497830.

44. Tudu, J.J., Panda, S.N., & Kumar, P. (2021). A comparative evaluation of process parameter optimization of wire cut electric discharge machining of Ti-6Al-4V using WASPAS and metaheuristic methods. *Psychology and Education Journal*, 58(1), 5860–5863.

45. Sahoo, P., Satpathy, M.P., Singh, V.K., & Bandyopadhyay, A. (2018). Performance evaluation in CNC turning of AA6063-T6 alloy using WASPAS approach. *World Journal of Engineering*, 15(6), 700–709.

46. Yazır, D., Şahin, B., & Yip, T. L. (2021). Selection of new design gas carriers by using fuzzy EVAMIX method. *The Asian Journal of Shipping and Logistics*, 37(1), 91–104.

47. Alinezhad, A., & Khalili, J. (2019). *New methods and applications in multiple attribute decision making (MADM)* (Vol. 277). Cham: Springer.

48. Veera P.D., & Ravipudi V.R. (2013). Application of AHP/EVAMIX method for decision making in the industrial environment. *American Journal of Operations Research*, 3(6), 542–569.

49. Della Spina, L. (2020). Adaptive sustainable reuse for cultural heritage: A multiple criteria decision aiding approach supporting urban development processes. *Sustainability*, 12(4), 1363.

50. Bandyopadhyay, S. (2021). Comparison among multi-criteria decision analysis techniques: A novel method. *Progress in Artificial Intelligence*, 10(2), 195–216.

51. Chatterjee, P., Athawale, V.M., & Chakraborty, S. (2011). Materials selection using complex proportional assessment and evaluation of mixed data methods. *Materials & Design*, 32(2), 851–860.

52. Chatterjee, P., & Chakraborty, S. (2013). Nontraditional machining processes selection using evaluation of mixed data method. *The International Journal of Advanced Manufacturing Technology*, 68(5–8), 1613–1626.

53. Baloyi, V.D., & Meyer, L.D. (2020). The development of a mining method selection model through a detailed assessment of multi-criteria decision methods. *Results in Engineering*, 8, 100172.

54. Madić, M., Petković, D., & Radovanović, M. (2015). Selection of non-conventional machining processes using the OCRA method. *Serbian Journal of Management*, 10(1), 61–73.

55. Ulutas, A., & Ölmez, U. (2019). Çok Kriterli Karar Verme Yöntemleri ile Kesintisiz Güç Kaynağı Seçimi. *Business & Organization Research (BOR) Conference*, Izmir, Turkey (pp. 1682–1689).

56. Chakladar, N.D., & Chakraborty, S. (2008). A combined TOPSIS-AHP-method-based approach for non-traditional machining processes selection. *Proceedings of the Institution of Mechanical Engineers, Part B: Journal of Engineering Manufacture*, 222(12), 1613–1623.

57. Chakladar, N.D., Das, R., & Chakraborty, S. (2009). A digraph-based expert system for non-traditional machining processes selection. *The International Journal of Advanced Manufacturing Technology*, 43(3), 226–237.

58. Rohith, R., Shreyas, B.K., Kartikgeyan, S., Sachin, B.A., Umesha, K.R., & Nanjundeswaraswamy, T.S. (2019). Selection of non-traditional machining process. *International Journal of Engineering Research & Technology*, 8(11), 148–155.

59. Patel, J.D., & Maniya, K.D. (2021). *Optimization of WEDM process parameters for aluminium metal matrix material Al+ SiC using MCDM methods*. In Advances in Manufacturing Processes (pp. 59–70). Singapore: Springer.

60. Kumar, R., & Ray, A. (2015). Optimal selection of material: an eclectic decision. *Journal of The Institution of Engineers (India): Series C*, 96(1), 29–33.

61. Patel, J.D., & Maniya, K.D. (2019). WEDM process parameter selection using preference ranking method: A comparative study. *International Journal of Manufacturing Research*, 14(2), 118–144.

62. Wang, J., Wei, G., Wei, C., & Wei, Y. (2020). MABAC method for multiple attribute group decision making under q-rung orthopair fuzzy environment. *Defence Technology*, 16(1), 208–216.

63. Božanić, D.I., Pamučar, D.S., & Karović, S.M. (2016). Application the MABAC method in support of decision-making on the use of force in a defensive operation. *Tehnika*, 71(1), 129–136.

64. Shivakoti, I., Peshwani, B., Kibria, G., Pradhan, B.B., & Sharma, A. (2019). Parametric optimization of WEDM using MABAC method. *8th International Conference on Modeling Simulation and Applied Optimization (ICMSAO)*, IEEE, Manama, Bahrain (pp. 1–4).

65. Zhang, S., Wei, G., Alsaadi, F.E., Hayat, T., Wei, C., & Zhang, Z. (2020). MABAC method for multiple attribute group decision making under picture 2-tuple linguistic environment. *Soft Computing*, 24(8), 5819–5829.

66. Pamucar, D., & Cirovic, G. (2015). The selection of transport and handling resources in logistics centers using Multi-Attributive Border Approximation area Comparison (MABAC). *Expert Systems with Applications*, 42(6), 3016–3028.
67. Alinezhad, A., & Khalili, J. (2019). *New methods and applications in multiple attribute decision making (MADM)* (Vol. 277). Cham: Springer.
68. Paramasivam, S.S.S.S., Kumaran, D., Natarajan, H., Krishnan, G.S., & Sairaghav, S.E. (2021). Process parameter optimization of key machining parameters of mg alloy with cryogenic treated tools by MABAC approach. *Materials Today: Proceedings*, 47, 7149–7154.
69. Yurdakul, M., & Cogun, C. (2003). Development of a multi-attribute selection procedure for non-traditional machining processes. *Proceedings of the Institution of Mechanical Engineers, Part B: Journal of Engineering Manufacture*, 217(7), 993–1009.
70. Chatterjee, P., Mondal, S., Boral, S., Banerjee, A., & Chakraborty, S. (2017). A novel hybrid method for non-traditional machining process selection using factor relationship and Multi-Attributive Border Approximation Method. *Facta Universitatis, Series: Mechanical Engineering*, 15(3), 439–456.

4 Taguchi-Based GRA Method for Multi-Response Optimization of Spool Bore in EHSV Made Up of Stainless Steel 440C

Pranav R.[1], Md. I. Equbal[2],
Azhar Equbal[3], and Kishore K.[4]
[1]Maturi Venkata Subba Rao Engineering College,
Hyderabad , India
[2]Mechanical Engineering Section, University Polytechnic,
 Aligarh Muslim University, Aligarh, U.P., India
[3]Faculty of Engineering and Technology,
 Department of Mechanical Engineering,
 Jamia Millia Islamia, New Delhi, India
[4]Vasavi College of Engineering, Ibrahim Bagh,
Hyderabad, India

LIST OF ACRONYMS

EHSV	Electro-hydraulic servo valve
WEDM	Wire electrical discharge machining
MRR	Material remove rate
Cyl.	Cylindricity
Ra	Surface roughness
ANOVA	Analysis of variance
GRG	Grey relational grade
GRA	Grey relational analysis
OAs	Orthogonal arrays
DOF	Degree of freedom
S/N	Signal-to-noise ratio

DOI: 10.1201/9780367822385-4

CONTENTS

4.1 Introduction 84
4.2 Experimental Details...85
    4.2.1 Selection of Processing Parameters..86
    4.2.2 Machining Profile 87
    4.2.3 Modules of Data Collection..88
    4.2.4 Design of Experiment...90
4.3 Methodology...90
    4.3.1 Grey Relational Generation 91
    4.3.2 Calculation of Grey Relational Grade (G$_i$) 92
4.4 Results and Analysis..93
    4.4.1 Multi-Response Optimization 96
4.5 Confirmation Tests...97
4.6 Conclusions...99
References...99

4.1 INTRODUCTION

Depending on the industrial application and service requirement, die sinking electrical discharge machining (EDM) and wire electrical discharge machining (WEDM) are the two major variants of EDM. Material removal in EDM machining is done by melting, vaporizing and controlled erosion by continuous electric sparks between workpiece and tool immersed in a dielectric medium [1]. Unlike the die sinking electrode, metal wire is used as the electrode in WEDM [1]. Commonly used wire electrodes are made from brass and coated steel wires, however, tungsten or molybdenum wires are preferred when a thin wire is required for machining [2]. During machining, two guide wires provided on above and below the workpiece are used to control the longitudinal movement of the wire. Wires ranging from 0.05 to 0.30 mm in diameter are used, which helps to cut difficult contours in hard-to-machine surfaces. A pulsed DC supply is used to generate a potential difference between the electrode and the workpiece [3]. The sparking between electrode and workpiece leads to the generation of heat, which results in the removal of material in the form of debris. Material is removed both from wire electrode and workpiece. WEDM machine also uses a nozzle that functions by injecting the coolant, commonly referred as dielectrics, into the machining zone and flushing away the debris. A computer-based positioning system maintains a constant gap between the wire electrode and the workpiece to be machined. WEDM is especially used in machining area, where close precision and very fine machined surface is a prerequisite. Different areas of application are in aerospace, medical, semiconductor, tool and die making and micro tooling. An electro-hydraulic servo valve (EHSV) is a major component of servo systems due to high degree of control precision, high reputability and accuracy, less weight, quick response and immunity from variations in load. Despite of many advantages offered by EHSV, it is a failure prone component that affects the reliability of a servo control system. Since inception, the technology of WEDM has undergone significant improvements to achieve the manufacturing requirements, and thus WEDM has developed as a suitable method for machining of EHSV [4]. To achieve a high rate of production in WEDM, economics of machining and optimization of processing parameter is very necessary. The study presents an experimental analysis with the aim of optimizing the WEDM process for machining of spool bore in EHSV. Optimization is proposed to control the performance measures, namely *MRR, Cyl.* and *Ra*. In a WEDM operation of spool bore, *MRR* determines the rate of production; *Ra* shows the machining quality and *Cyl.* is a measure of the deviation of a machined surface from that of a perfect cylinder [5]. Selections of optimal processing parameters have been investigated by many researchers, but the studies were mainly concentrated toward optimization of *MRR, Ra* and kerf width [5–10]. However, the major problem lies in the maintaining of cylindricity and smooth surface finish for maintaining the accuracy and reliability of the system. Thus, the following research gaps were recognized: (i) Optimization mainly focused toward improvement in *MRR, Ra* and kerf width (ii) Being a critical response affecting the system performance and reliability, cylindricity (*Cyl.*) is not valued by previous researchers. Thus authors got motivated to bridge this gap by considering the cylindricity of spool bore and its smooth surface finish as their prime performance output. The economy of machining is also considered by adding *MRR* into the list of investigating responses.

Taguchi-based orthogonal arrays (OAs) method is used for designing the experiments [11]. The predominant advantage of this technique lies in its simplicity and adaptability. They provide the required information making use of only the minimum number of experiments and still give outcomes which have good accuracy and are reproducible. The present work uses the Taguchi technique to design the experiment and optimizes the parameter with respect to the quality measures considered one at a time. Grey-based Taguchi analysis is used as a multi-response optimization method for simultaneous optimization of input variables i.e. Average machine voltage (*A*), time interval between two pulses (*B*), frequency (*C*) and wire tension (*D*) on different quality measures, namely *MRR, Ra* and *Cyl.* Taguchi-grey relational analysis (GRA) uses simple calculations and smaller sample size without any typical sample distribution and yields the result which does not have the conflicting conclusions from the qualitative analysis [12].

4.2 EXPERIMENTAL DETAILS

An EHSV, shown in Figure 4.1, is an electrically operated valve which regulates the hydraulic fluid sent to the actuator. It also offers accurate positional control, velocity and pressure control, force with good post-movement damping characteristics. The body of the EHSV is made of stainless steel (SS) 440C, which possesses high carbon content, shows higher strength, modest corrosion resistance and also has good hardness and wear resistance property. Grade 440C can be post-heat-treated to achieve the maximum wear resistance, hardness and strength among all the stainless steel alloy families. Table 4.1 presents the composition of 440C graded stainless steel. WEDM of 440C graded stainless steel is done using a brass wire electrode of 0.25 mm diameter with a vertical arrangement. ROBOFIL 240CC 5 axis

FIGURE 4.1   Body of a Type II EHSV.

CNC WEDM machine, manufactured by Charmilles (Figure 4.2), was used to conduct the experimentation in accordance with Taguchi L_9 OA.

The machine setup has a work table, power supply system, a dielectric flushing system, a positioning control system and a wire drive system. The machine offers the option of either coaxial flushing of the dielectric or having the workpiece submerged within the dielectric. The present work uses the submerged mode of arrangement as it ensures greater accuracy and better surface finish because of better cleaning of the debris provided in the gap between the workpiece and wire electrode and also better thermal stability. The process parameters are varied using the control panel.

4.2.1   SELECTION OF PROCESSING PARAMETERS

The parameters which were kept constant throughout the experimentation are referred as fixed parameters and are shown in Table 4.2. The process parameters were selected based on the preset arrangement of the settings for the production and process and from conducting the pilot experiments by varying one factor at a time. The process parameters taken into account are, namely, set value of average machine voltage (A), interval between two pulses (B), frequency (C) and wire tension (D).

Input parameters and their corresponding levels are presented in Table 4.3. The levels are chosen based on the data collected by running pilot experiments at different

TABLE 4.1

Composition of 440C Graded Stainless Steel in Wt%

Fe	C	Cr	Mn	Si	Mo
79.15	1.10	17.00	1.00	1.00	0.75

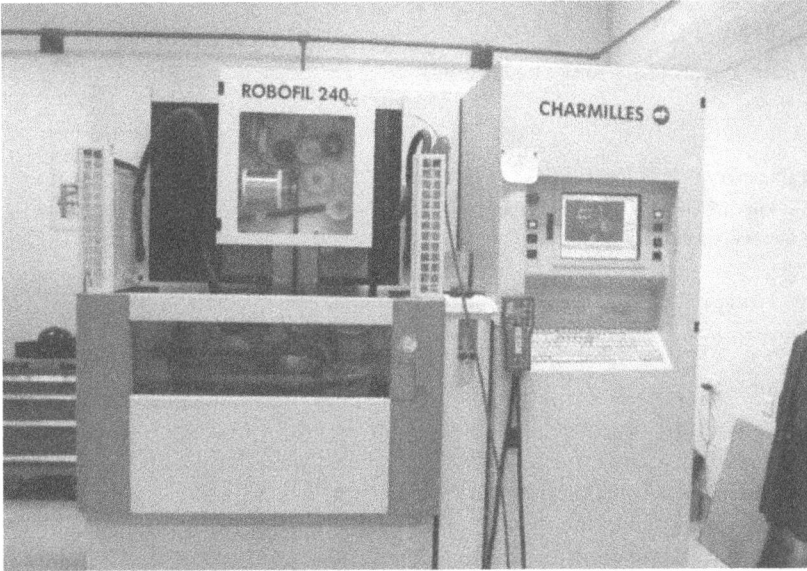

FIGURE 4.2 ROBOFIL 240CC WEDM setup.

values for the four process parameters varied one at a time and selecting those levels where an appreciable amount of change in the quality characteristics is observed.

4.2.2 Machining Profile

The machining process under investigation involves the enlargement of a spool bore from a given diameter of 4.00 mm to 4.480 mm. The cylindrical bore has a depth (height) of 30.50 mm. The Brass wire moves in a circular path following the inner periphery of the spool bore. During machining, the wire and the wall of the spool bore should be close enough without actually making any contact. The wire tool hence traces a circular path. A single experiment consists of 4 complete passes, two

TABLE 4.2
Fixed Parameters

Parameters	Value
Workpiece	Stainless steel 440C
Angle of cut	Vertical
Wire type	Brass (0.25 mm diameter)
Location of workpiece	Center of the table
Dielectric used	Water
Depth of workpiece	30.50 mm
Injection pressure	10 bar

TABLE 4.3

Input Parameters and Their Levels

Parameters	Unit	Symbol	Levels 1	2	3
Set value of average machine voltage	V	A	48.1	49.0	50.0
Time interval between two pulses	μs	B	6.50	7.00	7.50
Frequency	kHz	C	40.0	50.0	60.0
Wire tension	N	D	1.00	1.50	1.60

clockwise passes and two anticlockwise. The first run is anticlockwise, which performs a roughing operation, followed by three runs of finishing operations where the run occur in a sense that is clockwise, anticlockwise and clockwise, respectively. Numerical control part programming is generated in accordance with the defined wire path and fed into the machine through the control panel/computer. Required corrections can be made directly through the control panel on the machine or through the computer connected to the machine.

4.2.3   Modules of Data Collection

Material removal rate *(MRR)* – *MRR* is computed in accordance with the Eq. (4.1):

$$MRR = \frac{\text{Volume of the work piece lost during machining}}{\text{Time taken for machining}} \tag{4.1}$$

Workpiece dimensions are measured before and after machining using co-ordinate measuring machine (CMM) as shown in Figure 4.3. The difference in the volume of

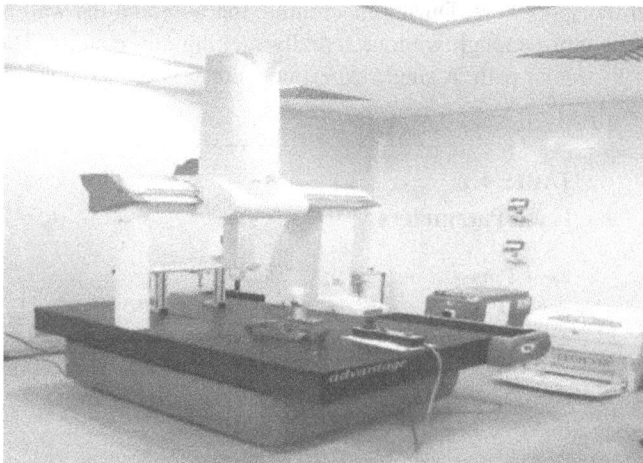

FIGURE 4.3   Hexagonal DEA CMM setup.

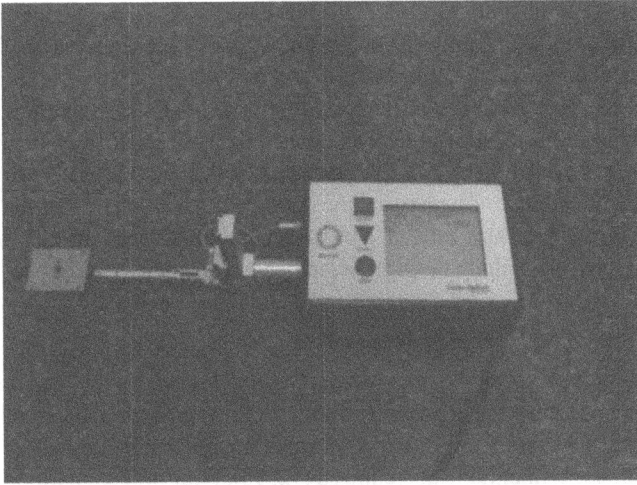

FIGURE 4.4 Taylor Hobson SURTRONIC 25.

the spool bore before and after machining is then divided with the total machining time to obtain the *MRR*. *MRR* is measured in mm^3/sec.

Surface roughness (*Ra*) – Taylor Hobson SURTRONIC 25 machine, as shown in Figure 4.4, is used to measure the average *Ra* value. A probe of diameter 4 mm is calibrated to a reference value of 6 microns using a reference plate (Figure 4.4). The traverse length is set as 30 mm and the three measurements taken are all along different traverse lines so as to not distort the succeeding reading due to the probe picking up any abnormality caused by the preceding traverse.

Cylindricity (*Cyl.*): Cylindricity of the hole is measured using a Taylor Hobson Talyrond 365 machine, as displayed in Figure 4.5. Eight reference circles at various

FIGURE 4.5 Taylor Hobson Talyrond 365 used to measure cylindricity.

heights in the spool hole are taken and the corresponding values are averaged to give us the average cylindricity value. The minimum zone reference circle method (LSCI) is used to calculate *Ra* value (Peak to valley out of roundness). In order to measure the roundness of the circle, rotation is coupled with the ability to measure the change in the radius. To achieve this profile of the component (spool bore) under testing is compared to a circular datum. Circular datum is provided by the rotation of the component on a highly accurate spindle. The axis of the spool bore is aligned with the axis of the spindle. As the component is rotated, a probe is entered from the top and measures the variations in the profile along the eight reference circles mentioned above.

4.2.4   DESIGN OF EXPERIMENT

Three quality characteristics, namely *MRR, Cyl.* and *Ra,* were evaluated. Experiments are designed in accordance with the full factorial design approach, which covers all the possible arrangements for a particular experimental setup. Taguchi's OA reduces the number of experimental trials required [12, 13]. Four process parameters each of three different levels are used, and thus each process parameter will contribute two degrees of freedom, thereby making total degree of freedom (DOF) as 8. Interactions among the parameters are neglected [14]. For the selection of OAs, DOF of the OAs must be equal to or greater than those of the process parameters. An L_9 array will have eight degrees of freedom (i.e.: $9 - 1 = 8$). It has already been specified that the process parameters used here have eight degrees of freedom. The DOF for the OAs is equal to the DOF of the process parameters, and hence L_9 for the experiments is chosen. Table 4.4 presents the experimental sequences in terms of an L_9 array considering the process parameters *A, B, C* and *D.*

4.3   METHODOLOGY

In accordance to Taguchi standard, S/N ratio signifies the signal-to-noise ratio where signal is the requisite value and the noise represents the unwanted value. This ratio also represents the ratio between mean and square deviation. It is designated by "*n*" and labeled in decibel (dB). Higher the better (HB) characteristic is chosen for *MRR,*

TABLE 4.4
Experimental Layout of L_9 Array

Exp. No.	A	B	C	D
1	48.1	6.5	40	1.0
2	48.1	7.0	50	1.5
3	48.1	7.5	60	1.6
4	49.0	6.5	50	1.6
5	49.0	7.0	60	1.0
6	49.0	7.5	40	1.5
7	50.0	6.5	60	1.5
8	50.0	7.0	40	1.6
9	50.0	7.5	50	1.0

whereas lower the better (LB) option is preferred for *Ra* and *Cyl*. S/N ratio for *MRR* is computed using Eq. (4.2) and *Ra* and *Cyl*. are calculated using Eq. (4.3).

$$\frac{S}{N} = -10\log_{10}\left(1/n\sum_{i=1}^{n}1/y_i^2\right); i = 1, 2..., N \tag{4.2}$$

$$\frac{S}{N} = -10\log_{10}\left(1/n\sum_{i=1}^{n}y_i^2\right); i = 1, 2,...., N \tag{4.3}$$

Taguchi method is preferred for optimization of a single objective function, but it cannot be implemented for optimization of multiple objective functions or responses directly. Hence, the data for every response observed using Taguchi's designs can be analyzed by other methods to obtain a solution for the multi-responses problems. The present research used Taguchi-based GRA method for multiple optimizations of the responses. The grey Taguchi technique is an approach for dealing with systems that are incomplete, uncertain or poor. The technique of GRA finds wide applications in processes with multiple responses or multiple quality characteristics such as rapid prototyping, EDM and welding [14–17]. GRA method translates multiple performance measures into a single response, namely grey relational grade (GRG) and GRG will help in determining the optimal combination of process parameters for all the responses simultaneously [18].

The steps used in the GRA method are:

  i. *Step* 1 – Depending on the nature of response i.e. either HB or LB, data obtained for quality characteristic is converted in terms of its S/N ratio (Y_{ij}) using the appropriate formula.
  ii. *Step* 2 – To reduce variability and to avoid the confusion of using different units for different quality characteristics Y_{ij} is normalized as Z_{ij} ($0 \leq Z_{IJ} \leq 1$). Normalizing is to transform the inputs, evenly distribute the data and scale it to an acceptable range.
  iii. *Step* 3 – From the normalized S/N ratios, grey relational coefficient (*GC*) is then calculated.
  iv. *Step* 4 – Grey relational grade (G_i) is computed.
  v. *Step* 5 – Optimal levels for the input variables or factors are then selected by using maximum average G_i values using the response graph method or ANOVA.

4.3.1 GREY RELATIONAL GENERATION

It is commonly observed that while measuring the performances of different quality characteristics, the influence of a few of them is neglected owing to the difference in the units of the characteristics. Sometimes this could also occur if some characteristics occur over a large range. This may also lead to incorrect result if the objective of quality characteristics is different. Hence, it becomes vital to normalize the data points, thereby processing the performance values for every alternative. This is called grey relational generation.

If the target value of the selected response is much larger, then "larger-the-better" characteristic is used and the results are expressed in accordance with Eq. (4.4). Here, in this study, this type of equation is used to normalize the values obtained for *MRR*.

$$Z_{ij} = \frac{Y_{ij} - \min(Y_{ij})}{\max(Y_{ij}) - \min(Y_{ij})} \tag{4.4}$$

The normalized smaller the better characteristic is expressed in accordance with Eq. (4.5). Eq. (4.5) is used to normalize the values obtained for *Cyl.* and *Ra*.

$$Z_{ij} = \frac{\max(Y_{ij}) - Y_{ij}}{\max(Y_{ij}) - \min(Y_{ij})} \tag{4.5}$$

After performing the grey relational generation operation, every responses are scaled into [0, 1], where closeness to 1 or equal to 1 is considered as the best alternative or vice-versa. While, alternative closest to or equal to 1 does not actually exist and hence reference sequence $Y_o = \{Y_{oj} = 1 | j = 1, 2, 3, ..., n\}$ is opted to suggest the alternate having comparability sequence closest to the reference sequence. For the same, grey relational coefficient (GC) is considered. For a larger value of GC Y_{ij} and Y_{oj} are closer. Eq. (4.6) is used to compute the GC,

$$GC_{ij} = \frac{\Delta_{\min} + \lambda \Delta_{\max}}{\Delta_{ij} + \lambda \Delta_{\max}}, \tag{4.6}$$

where Δ denotes the absolute difference between Y_{oj} and Y_{ij}, which shows its deviance from the target value and it is considered as quality loss. $\Delta_{\min}$ and $\Delta_{\max}$ refer to the minimum and the maximum values of the delta. Y_{oj} and Y_{ij} are the optimal performance or normalized value and the *i*th normalized value of the *j*th response variable.

λ denotes the distinguishing coefficient defined in the range $0 \le \lambda \le 1$. Here we have taken the value of λ to be 0.33 since we have three quality characteristics and we are giving the three of them equal weightage. λ is employed for expanding or compressing the range of grey relation coefficient [13]. λ in the defined range provides the equal design of levels of factor without disturbing them [13].

4.3.2   CALCULATION OF GREY RELATIONAL GRADE (G_i)

G_i is used to represent the relationship between the comparability sequence and the reference sequence and is computed using Eq. (4.7).

$$G_i = \frac{1}{m} \Sigma GC_{ij} \tag{4.7}$$

Higher value of G_i represents the closer corresponding combination of factors to the optimal value. G_i is considered as a single response problem and values are analyzed to decide the optimal factors and their corresponding levels.

TABLE 4.5
Experimental Results

Exp. No.	MRR (mm³/s)	S/N ratio for MRR (dB)	Cyl. (μm)	S/N ratio for Cyl. (dB)	Ra (μm)	S/N ratio for Ra (dB)
1	0.089773	−20.9371	4.71667	−13.4727	0.723333	2.81323
2	0.103887	−19.6688	5.22333	−14.3590	0.650000	3.74173
3	0.101803	−19.8448	5.67000	−15.0717	0.670000	3.47850
4	0.102590	−19.7779	5.31000	−14.5019	0.620000	4.15217
5	0.090570	−20.8603	5.38333	−14.6210	0.636667	3.92176
6	0.102977	−19.7452	4.92667	−13.8511	0.603333	4.38885
7	0.095690	−20.3827	5.58333	−14.9379	0.536667	5.40591
8	0.100597	−19.9483	4.97667	−13.9388	0.556667	5.08810
9	0.089033	−21.0089	5.06000	−14.0830	0.616667	4.19899

4.4 RESULTS AND ANALYSIS

The experiments are conducted following the OA given in Table 4.4. Three sets of experiments are conducted for each run, and the average values of the quality measures are calculated and tabulated. Table 4.5 shows the results from experimentation. The S/N ratios for *MRR* are calculated using Eq. (4.2) and *Cyl.* and *Ra* are computed in accordance with Eq. (4.3).

Once the calculations are made, the graphs for the particular control parameters at their three levels of application are plotted. Figures 4.6–4.8 show the graphs obtained for S/N ratios at different levels of application. The maxima of the main effects plot of each control parameter correspond to the optimal level of

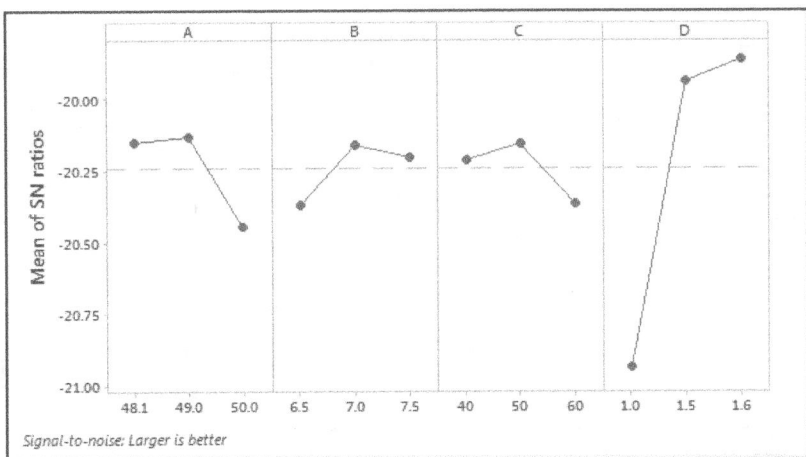

FIGURE 4.6 Main effects plot for S/N ratios of *MRR*.

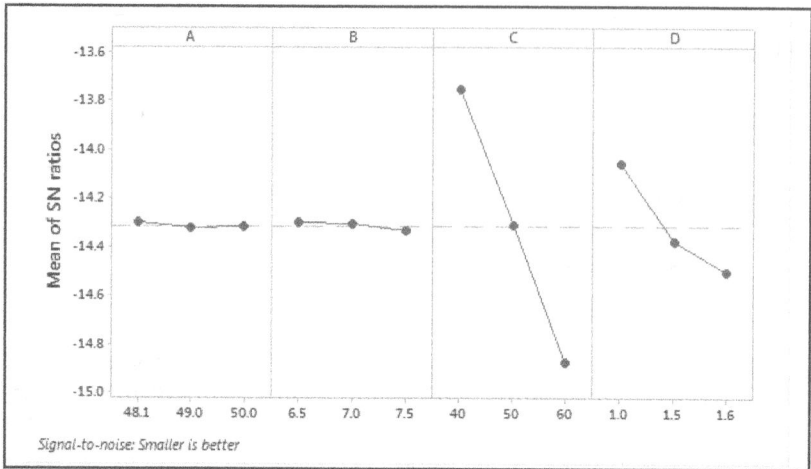

FIGURE 4.7 Main effect plot of S/N ratio for *Cyl.*

the levels so chosen during the design of experiment for that quality characteristic. Table 4.6 gives the optimal levels of factors achieved using main effect plots shown in Figures 4.6–4.8. The relative significance of each factor is established using ANOVA and Tables 4.7– 4.9 present the result conforming to each studied response [19, 20]. Here, the degree of freedom is represented by *DOF*, *SS* represents the sum of square, Variance is denoted by v and percentage contribution by (% *C*). It is observed from Table 4.7 that *D* and *A* are the two main factors for determining the material removal of the spool bore (SS 440C) by WEDM. Analysis of results showed that machining speed is inversely proportional to the set voltage

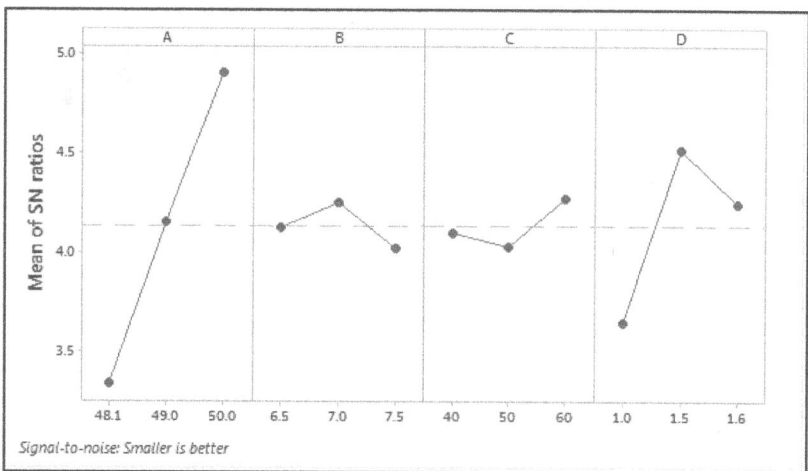

FIGURE 4.8 Main effect plot for S/N ratio of *Ra.*

TABLE 4.6
Optimal Levels of Process Parameters

Process Parameter	Optimal Level		
	MRR	*Cyl.*	*Ra*
A	49.0	48.1	50.0
B	7.00	6.50	7.00
C	50.0	40.0	60.0
D	1.60	1.00	1.50

value. Ionization takes place in the dielectric with the decrease in the voltage value hence causing more erosion of the workpiece, whereas a larger wire tension leads to a quick escape of the eroded material from within the spark gap, which aids in an increased *MRR*. Table 4.8 proposes that the frequency of the pulses (*C*) and the wire tension (*D*) play a major role in obtaining the best cylindrical bores. As *C* is increased, the energy available to erode the workpiece is distributed over a larger number of sparks, and hence the craters formed due to the erosion are reduced in size, which not only leads to a better surface finish but also helps maintain the machining profile better hence helping to maintain better cylindricity. The change in the tension of the wire may lead to varying overcuts and undercuts that affect the circularity for the difference reference circles taken, which in turn affect the cylindricity. Machine voltage (*V*) and wire tension (*D*) significantly affect the *Ra* value of the machined surface, as depicted in Table 4.9.

Increase in *D* reduces the vibrations that are caused in the wire and it can be observed that the direction in which the wire is fed has lesser variations in vibrations giving rise to a better surface finish. Increase in *V* also shows an increase in *Ra* that can be attributed to the craters created due to erosion being larger in size.

TABLE 4.7
ANOVA for *MRR*

Parameters	DOF	SS	V	% C	F-value
A	2	2.45×10^{-5}	1.22×10^{-5}	7.96748	1.282723
B	–	9.3×10^{-6a}	–	–	–
C	–	9.8×10^{-6a}	–	–	–
D	2	0.000264	0.000132	85.82114	13.81675
Error	4	0.0000191	–		
Total	8	0.000308			

[a] *Pooled*

TABLE 4.8
ANOVA for *Cyl.*

Parameters	DOF	SS	V	% C	F-value
A	–	0.000022[a]	–	–	–
B	–	0.000919[a]	–	–	–
C	2	0.678085	0.339043	85.65876	720.6004
D	2	0.112586	0.056293	14.22237	119.6451
Error	4	0.000941	–		
Total	8	0.791612			

[a] *Pooled*

4.4.1 MULTI-RESPONSE OPTIMIZATION

Table 4.6 reveals that the optimal factor levels for *MRR*, *Cyl.* and *Ra* are different. ANOVA tables, as presented in Tables 4.7–4.9 for individual responses, also present the significant factors are also different for all the responses studied. As the industrial requirement needs only one factor setting at which part must be machined to yield optimum responses and thus multi-response optimization technique is needed to propose one optimal factor setting at which WEDM should be performed. Since it has already been discussed in the chapter that Taguchi method alone is not capable of solving problems involving multiple responses. Hence, the present study makes use of a Taguchi technique in combination with GRA method for the optimization of multiple responses (*MRR*, *Ra* and *Cyl.*). The present work uses L_9 OA for designing of experimental sequence and GC_{ij} and G_i for each experiment performed is presented in Table 4.10. Figure 4.9 shows the main effect plot for G_i, which proposes the factors and levels: A_1, B_3, C_3 and D_3 as the optimal factor setting. ANOVA analysis performed on G_i, as given in Table 4.11, suggest that factor *A* and *D* are significant.

TABLE 4.9
ANOVA for *Ra*

Parameters	DOF	SS	V	% C	F-value
A	2	0.01858	0.00929	72.93063	23.51614
B	–	0.000403[a]	0.000201	1.579894	–
C	–	0.000388[a]	0.000194	1.521408	–
D	2	0.006106	0.003053	23.96806	7.728389
Error	4	0.00079	–		
Total	8	0.025476			

[a] *Pooled*

TABLE 4.10
Calculation of Grey Relational Grade (*Gi*)

Exp. No.	Normalized S/N ratios (Z_{ij})			Quality Loss Function (Δ)			Grey Relational Coefficient (GC_{ij})			Grey Relational Grade (G_i)
	MRR	Cyl.	Ra	MRR	Cyl.	Ra	MRR	Cyl.	Ra	
1	0.05358	0.00000	1.00000	0.94642	1.000000	0.000000	0.25854	0.24812	1.00000	0.50222000
2	1.00000	0.55426	0.64188	0.00000	0.445716	0.358124	1.00000	0.42540	0.47956	0.63498667
3	0.86867	1.00000	0.74340	0.13133	0.000000	0.256595	0.71532	1.00000	0.56257	0.75929667
4	0.91859	0.64365	0.48357	0.08141	0.356348	0.516431	0.80212	0.48080	0.38987	0.55759667
5	0.11089	0.71814	0.57244	0.88911	0.281864	0.427561	0.27069	0.53934	0.43561	0.41521333
6	0.94299	0.23665	0.39228	0.05701	0.763352	0.607719	0.85269	0.30182	0.35192	0.50214333
7	0.46728	0.91632	0.00000	0.53272	0.083677	1.000000	0.38251	0.79772	0.24812	0.47611667
8	0.79143	0.29149	0.12258	0.20857	0.708505	0.87742	0.61273	0.31776	0.27331	0.40126667
9	0.00000	0.38168	0.46551	1.00000	0.618324	0.534489	0.24812	0.34798	0.38173	0.32594333

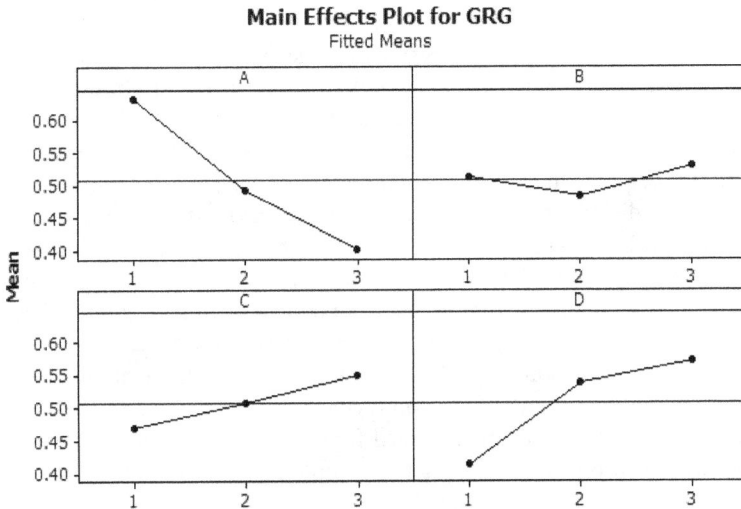

Main Effects Plot for GRG
Fitted Means

FIGURE 4.9 Main effect plot for G_i.

4.5 CONFIRMATION TESTS

To verify and validate the experimental conclusions, confirmation experiments are conducted by determining the results of the test using a specific combination of the factors and levels. Eq. (4.8) is used to predict and verify the experimental conclusion.

$$\eta_{opt} = \eta_m + \sum_{j=1}^{k}(\eta_j - \eta_m), \qquad (4.8)$$

TABLE 4.11
ANOVA Table for *Gi*

Parameters	DOF	SS	v	% C	F-value
A	2	0.08133	0.040665	59.81686	25.90953
B	–	0.003139[a]	0.00157	2.308682	–
C	2	0.010024	0.005012	7.372486	3.193374
D	2	0.041471	0.020736	30.50123	13.21153
Error	2	–	–		
Total	8	0.135965			

[a] *Pooled*

where η_{opt} is grand mean of S/N ratio, η_j is mean S/N ratio at optimum level and k is the number of main input parameters affecting the responses.

Using the relationship given in Eq. (4.8), we calculate the theoretical values for the S/N ratios for the different responses at their optimum level. The results are correlated with the values obtained by performing the experiments at the optimized levels and are tabulated in Table 4.12. At optimal factor setting for different individual responses, experiments were conducted and the result obtained was compared with the value achieved using the predictive confirmation equation as given in Eq. (4.8).

Based on the percentage error as obtained in Table 4.12, the resulting model is expected to predict the reasonable value of *MRR*, *Cyl*. And *Ra*. Error % of 0.48, 0.18 and 6.78 are observed for the S/N ratio of *MRR*, *Cyl*. and *Ra*, which shows the predicting accuracy of the proposed model. Small error percentage of 7.64 between the experimental and predictive value of G_i proves the appropriateness of the resulting model for multiple responses during WEDM of stainless steel 440C, which is used in spool bore type II EHSV. It is important to mention that increase in the number of experimental runs can further reduce the error and can be treated as a future work. This confirms the need for a mathematical model for predicting the measures of responses according to the input parameter assigned.

TABLE 4.12
Comparison of S/N Ratios between Experimental and Predicted Optimized Results

Responses	Optimized Setup	Predicted S/N Ratio (in dB)	Experimental S/N Ratio (dB)	Error (%)
MRR	$A_2B_2C_2D_3$	−19.7432	−19.6492	0.48
Cyl.	$A_1B_1C_1D_1$	−13.4979	−13.4727	0.18
Ra	$A_3B_2C_3D_2$	5.27769	5.66140	6.78
G_i	$A_1B_3C_3D_3$	0.701227	0.759297	7.64

4.6   CONCLUSIONS

An experimental investigation was performed to obtain an optimal combination of process parameters for the WEDM of a spool bore for a type II EHSV. The spool bore is made by machining of stainless steel of grade 440C. Average machine voltage (V), interval between two pulses (B), frequency (C) and wire tension (D) were chosen as the driving parameters and material removal rate (MRR), cylindricity ($Cyl.$) and surface roughness (Ra) were recorded as the output response. Experiments were conducted in accordance with Taguchi L_9 array. Taguchi-based GRA method is employed as a multi-response optimization method to simultaneously optimize the MRR, $Cyl.$ and Ra. Key conclusions drawn from the investigation are:

1. For optimum MRR, the recommended parametric combination is $A_2B_2C_2D_3$, where A_2 is 49 V, B_2 is 7 μs, C_2 is 50 kHz and D_2 is 1.5 N. For better cylindricity ($Cyl.$), optimal parametric combination is $A_1B_1C_1D_1$, where A_1 is 48.1 V, B_1 is 6.5 μs, C_1 is 40 kHz and D_1 is 1 N. For optimal Ra the recommended parametric combination is $A_3B_2C_3D_2$, where A_3 is 50 V, B_2 is 7 μs, C_3 is 60 kHz and D_2 is 1.5 N.
2. Optimization of machining process revealed that A and D are the key factors affecting the MRR. Frequency of the pulses (C) and the wire tension (D) are important to achieve the best cylindrical bores. Ra of the machined SS 440C spool bore is primarily influenced by A and D during WEDM operation.
3. Interval between two pulses (B) is not found to be significantly affecting either of the performance measure studied.
4. Multi-response optimization was performed for all the chosen responses by using grey Taguchi method and the optimal factors and levels were determined. Optimal parametric combination found is $A_1B_3C_3D_3$.
5. ANOVA analysis of G_i proposes A and D as the major factors affecting the WEDM process of spool bore for a type II EHSV made from stainless steel 440C.
6. To validate the efficacy of the developed model, confirmation experiments were conducted and the results were compared with experimental results. The small error of magnitude 0.48%, 0.18%, 6.78% and 7.64% between the experimental result and model result is observed for MRR, $Cyl.$, Ra and G_i, respectively, which shows the developed model is capable of predicting significant results.

REFERENCES

1. M. Durairaj, D. Sudharsun and N. Swamynathan, Analysis of process parameters in wire EDM with stainless steel using single objective Taguchi method and multi objective grey relational grade, Procedia Engineering, 64, 868–877, 2013.
2. J. Kapoor, S. Singh and J. S. Khamba, Recent developments in wire electrodes for high performance WEDM, Proceedings of the World Congress on Engineering, 2, 1065–1068, 2010.
3. D. Ghodsiyeh, A. Golshan and A. S. Jamal, Review on current research trends in wire electrical discharge machining (WEDM), Indian Journal of Science and Technology, 6(2), 4128–4140, 2013.

4. D. T. Pham, S. S. Dimov, S. Bigot, A. Ivanov and K. Popov, Micro EDM-recent developments and research issues, Journal of Materials Processing Technology, 149(1–3), 50–57, 2004.

5. S. S. Mahapatra and A. Patnaik, Optimization of wire electrical discharge machining (WEDM) process parameters using Taguchi method, International Journal of Advanced Manufacturing Technology, 34(9–10), 911–925, 2007.

6. Z. A. Khan, A. N. Siddiquee, N. Z. Khan, U. Khan and G. A. Quadir, Multi response optimization of wire electrical discharge machining process parameters using Taguchi based Grey relational analysis, Procedia Material Science, 6, 1683–1695, 2014.

7. S. Tilekar, S. S Das and P. K. Patowarib, Process parameter optimization of Wire EDM on Aluminum and mild steel by using Taguchi method, Procedia Materials Science, 5, 2577–2584, 2014.

8. J. U. Prakash, S. J. Juliyana, S. Karthik and T. V. Moorthy, Optimization of wire EDM process parameters for machining of AMCs (413/B_4C) using Taguchi technique, International Journal of Mechanical and Production Engineering Research and Development, 7(6), 231–238, 2017.

9. J. U. Prakash, S. J. Juliyana, P. Pallavi and T. V. Moorthy, Optimization of Wire EDM process parameters for machining hybrid composites (356/B_4C/Fly Ash) using Taguchi technique, Materials Today: Proceedings, 5(2), 7275–7283, 2018.

10. R. Magabe, N. Sharma, K. Gupta and J. P. Davim, Modeling and optimization of Wire-EDM parameters for machining of $Ni_{55.8}$Ti shape memory alloy using hybrid approach of Taguchi and NSGA-II, The International Journal of Advanced Manufacturing Technology, 102(5–8), 1703–1717, 2019.

11. S. D. Bolboacă and L. Jäntschi, Design of experiments: Useful orthogonal arrays for number of experiments from 4 to 16, Entropy, 9, 198–232, 2007.

12. A. K. Sood, R. K. Ohdar and S. S. Mahapatra, Improving dimensional accuracy of fused deposition modelling processed part using grey Taguchi method, Materials & Design, 30(9),4243–4252, 2009.

13. D. C Montgomery, Design and Analysis of Experiments, John Wiley and Sons, Eighth Edition, Hoboken, 2013.

14. K. Jangraa, S. Grovera and A. Aggarwal, Simultaneous optimization of material removal rate and surface roughness for WEDM of WC-Co composite using grey relational analysis along with Taguchi method, International Journal of Industrial Engineering Computations, 2, 479–490, 2011.

15. M. Lu and K. Wevers, Grey system analysis and applications: A way forward, Journal of Grey System, 10(1), 47–54, 2007.

16. P. Sathiya and M. Y. Abdul Jaleel, Grey based Taguchi method for optimization of bead geometry in laser bead on plate welding, Advances in Production Engineering and Management, 5(4), 225–234, 2010.

17. Y. S. Tarng, S. C. Juang and C. H. Chang, The use of grey based Taguchi methods to determine submerged arc welding process parameters in hard facing, Journal of Material Processing Technology, 128, 1–6, 2002.

18. Y. Kuo, T. Yang and G. W. Huang, The use of grey-based Taguchi method to optimize multi response simulation problems, Engineering Optimization, 40(6), 517–528, 2008.

19. A. Equbal, M. I. Equbal and A. K. Sood, An investigation on the feasibility of fused deposition modelling process in EDM electrode manufacturing, CIRP Journal of Manufacturing Science and Technology, 26, 10–25, 2019.

20. A. Equbal, M. I. Equbal and A. K. Sood, PCA-based desirability method for dimensional improvement of part extruded by fused deposition modelling technology, Progress in Additive Manufacturing, 4(3), 269–280, 2019.

5 Wearing Behaviour of Electrodes during EDM of AISI 1035 Steel

Azhar Equbal[1] *and Md. Asif Equbal*[2]
[1]Faculty of Engineering and Technology, Jamia Millia
Islamia (A Central University), New Delhi, India
[2]Cambridge Institute of Technology, Ranchi, India

CONTENTS

5.1   Introduction 101
5.2   Methodology 103
5.3   Results and Discussion 104
5.4   Optimization 106
5.5   Conclusions 109
References 110

5.1   INTRODUCTION

Electrical discharge machining (EDM) is a non-traditional machining process that uses energy from repetitive sparks occurring between tool and workpiece to finish the work material [1–3]. Two main components in EDM are workpiece and tool, where the tool is also referred as an electrode. The workpiece is clamped and fixed in a fixture and the electrode is held in a tool holder. The schematic diagram explaining the machining process is shown in Figure 5.1. The entire process of machining takes place inside an insulating medium which is industrially known as dielectric [4]. This dielectric helps in maintaining a plasma channel between the electrode and workpiece. The sudden collapse of the plasma channel creates a series of sparks between electrode and workpiece which promotes the machining of workpiece [5]. A precise gap is maintained between the electrode and workpiece using a servo controller to help the timely occurrence of the spark. The most common dielectric used in industrial applications is kerosene, but other hydrocarbon-based oils and silicon-based oils are also used. Recent advances, research, and development in the field showed that deionized water and air can also be used as dielectric fluid [6].

For a workpiece, machining performances are measured in terms of material removal rate (*MRR*), surface roughness (*SR*), and dimensional accuracy (*DA*) [7]. For the determination of *MRR*, the weight of the workpiece before and after machining is recorded (for each cavity), and the difference in weight is divided by

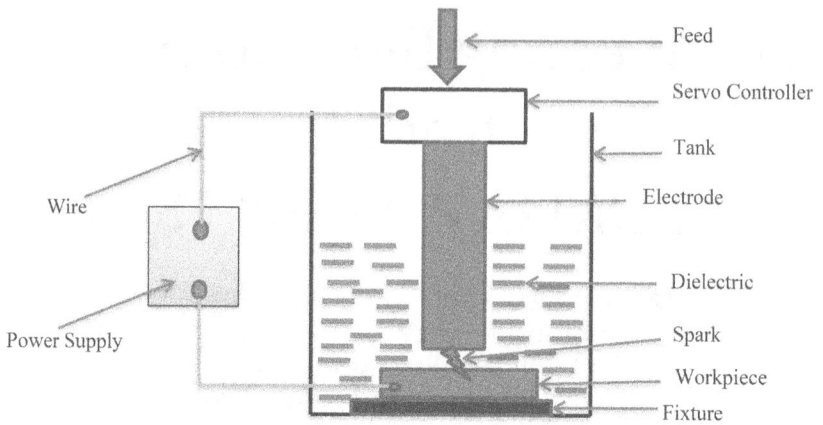

FIGURE 5.1   Schematic diagram showing mechanism in EDM.

the time of machining for that particular cavity. It is generally expressed in grams per minute (g/min) [7]. *SR* is determined by roughness testing machines or using coordinate measuring machines. *SR* is generally expressed in micrometre (μm). The *DA* of the machined cavity is calculated by finding the deviations in machined diameter from the original diameter (ΔD) and deviations in machined depth from the original depth (ΔH) [7]. Since repetitive sparks not only result in finishing of workpiece but in addition undesirable wearing of electrode also occur which makes tool wear ratio (*TWR*) as an important performance measure. Since, minimum *TWR* is important in EDM, it is necessary to understand the complex wearing behaviour of electrodes obtained during machining. Wear variation in electrodes was determined by Mohri et al. [8] for different electrode profiles in EDM. They observed that the edge of the electrodes adds to the maximum electrode wear. They also proposed that major electrode wear take place at the beginning of machining. As the machining progresses, a black layer of carbon gets deposited on the surface of the electrode, which minimizes the electrode wear. They also observed that electrode wear is significantly affected by the chosen machining factors and witnessed zero electrode wear at a longer pulse on time. A study on wearing of electrode was also performed by Pham et al. [9] during micro-EDM. A method for calculation of volumetric wear ratio was suggested. They mentioned that wear is dominated by machining factors proposed and wear compensation method to minimize the wear of electrode. The wear compensation method was also suggested by Kar and Patowari [10] for less electrode wear during the micro-EDM. Khan [11] investigated the wear of copper and brass electrodes during EDM of aluminium and mild steel. It was noticed that electrode wear significantly increases with the rise in current and voltage. In addition, they concluded that wear observed along the cross section of electrode is higher than wear observed along the length of the electrode. Laurenţiu et al. [12] experimentally suggested that electrodes wear also depends on the material of electrode. They verified that minimal electrode wear was observed while using steels and aluminium in comparison to copper, which shows intense wear. Sltineanu et al. [13] investigated the wear of tool electrodes during EDM of smaller holes. They concluded that electrode wear significantly increases for a smaller diameter of electrode. Once the diameter of the working electrode increases, electrode wear decreases because of improved heat evacuation from the work zone or less heat transferred to the tool electrode. Equbal et al. [7] use the copper metallization technique over FDM (fused deposition modelling) fabricated parts to convert them into conductive tool electrodes. The use of FDM-based EDM electrode also reflected that its wear is mostly affected by chosen EDM factors. For minimizing the *TWR,* optimization of EDM factors was suggested. Heidari et al. [14] suggested that the wear resistance of copper electrode increases by the use of a very fine-grained structure.

Literature manifest that electrode wear in EDM is affected by a number of factors like electrode material, carbon deposition on electrode, size of hole machined, and chosen machining factors. While it is difficult to control the other parameters involved in machining, it is a wise decision to choose an optimal range of machining factors. To achieve a good electrode performance, minimum wear of electrode is desirable, and hence this chapter is aimed at optimization of EDM parameters to achieve minimum *TWR.* Current work undertaken the machining of AISI 1035 steel under the influence of three main EDM parameters, viz. current (I), pulse on time (T_{on}), and pulse off time (T_{off}). For a better understanding of the influence of these factors on *TWR,* they are varied at three levels each.

5.2 METHODOLOGY

Design of experimentation is done by using the full factorial design (FFD) technique [15]. FFD considers all the possible experimental runs combining factors and levels undertaken. The present study used three EDM factors, viz. current (I), pulse on time (T_{on}), and pulse off time (T_{off}), and AISI 1035 steel was chosen as the workpiece material. The cylindrical electrode of a diameter 5.8 mm was prepared and finished by machining on CNC lathe. Selected EDM factors and their levels are shown in Table 5.1.

TABLE 5.1
EDM Factors and Their Levels [16]

EDM Factors	Notation	Levels			Unit
		1	2	3	
		Low Level (1)	Centre Level (2)	High Level (3)	
Current	I	5	10	15	A
Pulse on time	T_{on}	150	300	450	μs
Pulse on time	T_{off}	90	150	210	μs

The factors and levels are selected in accordance with machining conditions used in EDM i.e. rough machining, semi-finish machining, and finish machining [16]. EDM was done using Vidyunt (MMT, ZNC) EDM machine, and data for tool wear rate (*TWR*) are collected as under:

Tool wear rate (*TWR*): Wears out material from the electrode was computed in g/min using Eq. (5.1).

$$TWR = \frac{m_i - m_f}{t}, \tag{5.1}$$

where, m_i and m_f are the electrode weight in grams prior to and post-machining measured using a weighing machine (Model: Mettler PM1200, Make: India) and t is the machining time (in minutes) noted down using the stopwatch of the mobile phone. For analysing the result of experimentation analysis of variance (ANOVA) technique is used [17]. ANOVA is to evaluate the influence of parameters and interactions on *TWR*. Significance factors and interactions were determined using a p-value where p value ≤ 0.05 is considered as significant for a significance level of 5%. The effectiveness of the model is established using the Anderson-Darling (A-D) plot. Further, the setting of the EDM factors for optimum *TWR* was demonstrated using the main effect plot. To verify the results, optimization is also done using the desirability function approach.

5.3 RESULTS AND DISCUSSION

The results of experimentation are tabulated in Table 5.2. Here, 1 corresponds to the lower level, 2 corresponds to the middle level, and 3 corresponds to the higher level as shown in Table 5.1. *TWR* is computed in g/min. ANOVA result for *TWR* is presented in Table 5.3.

Here, *SS* denotes the sum of square and *DF* means the degrees of freedom. As per ANOVA analysis, it was observed that I, T_{on} and $I \times T_{on}$ was significant. For the better understanding of the reader interaction plot is also presented in Figure 5.2, which also shows $I \times T_{on}$ was significant. A-D plot, as shown in Figure 5.3, shows that residuals are near to the centreline, and their deviations are within the control line. The regression equation for *TWR* is given in Eq. (5.2), which is valid at R^2 of 99%. The R^2 of 0.99 showed that the model defined in Eq. (5.2) is very accurate for the prediction of *TWR*.

$$TWR = -1.06678 + 0.796A + 0.396B - 0.005C$$
$$-0.287AB + 0.031AC - 0.017BC \tag{5.2}$$

ANOVA analysis showed that interaction $I \times T_{on}$ is significant, and hence 3D surface plot for $I \times T_{on}$ is presented in Figure 5.4. The surface plot presented in Figure 5.4 concludes that *TWR* increases with the rise in I. At higher I, the intensity of the spark is increased, which results in greater *TWR* [18]. However, the increase in tool wear at a lower value of T_{on} is more significant.

TABLE 5.2
EDM Experimental Result

	Factors			
Run	**Current (I)**	**Spark on Time (T_{on})**	**Spark off Time (T_{off})**	TWR **(g/min)**
1	1	1	1	0.012
2	1	1	2	0.011
3	1	1	3	0.012
4	1	2	1	0.008
5	1	2	2	0.004
6	1	2	3	0.005
7	1	3	1	0.001
8	1	3	2	0.001
9	1	3	3	0.001
10	2	1	1	0.104
11	2	1	2	0.154
12	2	1	3	0.125
13	2	2	1	0.047
14	2	2	2	0.049
15	2	2	3	0.067
16	2	3	1	0.016
17	2	3	2	0.039
18	2	3	3	0.035
19	3	1	1	1.142
20	3	1	2	1.335
21	3	1	3	1.443
22	3	2	1	0.278
23	3	2	2	0.287
24	3	2	3	0.245
25	3	3	1	0.118
26	3	3	2	0.105
27	3	3	3	0.221

At lower T_{on} more heat is transferred to the tool, and hence TWR is more. At higher T_{on} the diameter of spark is increased owing to which more heat is transferred to work, but heat transfer to the tool is decreased as only a portion of the tool bottom is exposed to heat and the remaining is lost to surrounding, thus resulting in low TWR [19]. Figure 5.5 showed that the effect of T_{off} on TWR is not significant, which can also be verified from the ANOVA table presented in Table 5.3. At low T_{off}, TWR increases marginally as the duration between successive sparks is less, but once the T_{off} is increased to a higher value, ample time is available for repossession of dielectric strength. Once the dielectric strength is regained, a major portion of discharge energy in the next cycle is used in overcoming that regained strength which results in minimal tool wear [20].

TABLE 5.3
ANOVA Table for *TWR*

Source	DF	Seq *SS*	Adj *SS*	Adj *MS*	F	p
I	2	1.74575	1.74575	0.87287	320.1	0.000
T_{on}	2	0.95785	0.95785	0.47892	175.63	0.000
T_{off}	2	0.01033	0.01033	0.00516	1.89	0.212
$I \times T_{on}$	4	1.4899	1.4899	0.37248	136.59	0.000
$I \times T_{off}$	4	0.01367	0.01367	0.00342	1.25	0.363
$T_{on} \times T_{off}$	4	0.01156	0.01156	0.00289	1.06	0.435
Error	8	0.02182	0.02182	0.00273		
Total	26	4.25087				
R^2	0.99					

5.4   OPTIMIZATION

To show the individual effect of chosen EDM factors on *TWR*, main effect plot is shown in Figure 5.6. In the main effect plot, the mean response of each level factor is linked by a line. A horizontal line demonstrates that no effect is present. A small deflection from a horizontal line minutely affects the response. The large slope of the line from a horizontal orientation demonstrates the greater magnitude of the main effect. Here, for a minimum *TWR* individual I (5A), large T_{on} (450 µs), and low

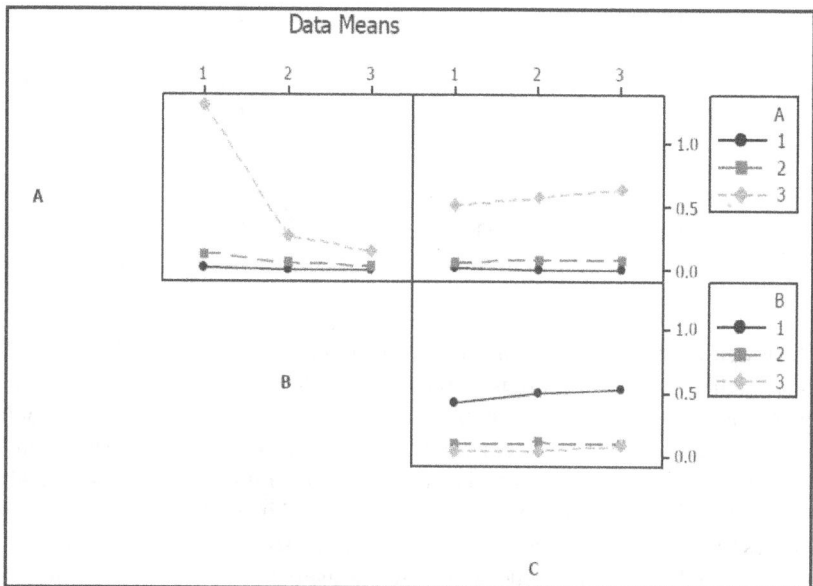

FIGURE 5.2   Interaction plot for *TWR*.

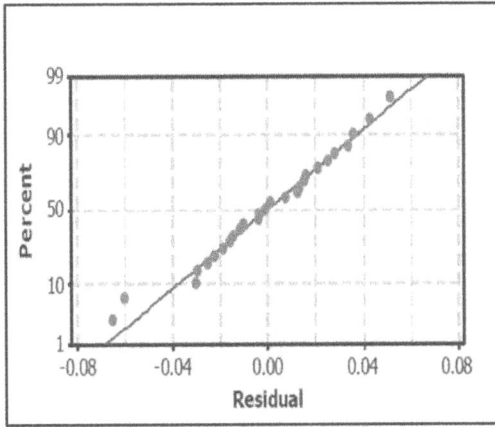

FIGURE 5.3 A-D plot for *TWR.*

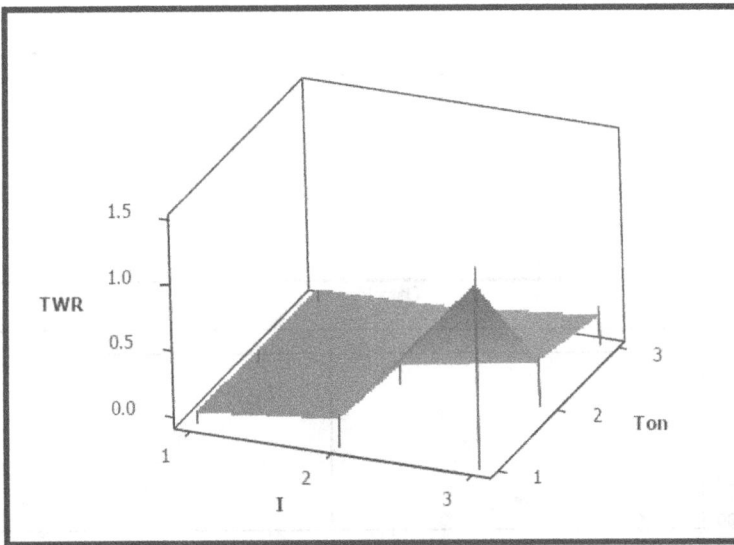

FIGURE 5.4 Surface plot for $I \times T_{on}$.

T_{off} (90 µs) are required. The optimal setting of EDM parameters which results in minimal *TWR,* is also verified using the desirability function approach. The result of desirability is shown in Figure 5.7. The optimization showed that minimal *TWR* is obtained at $I = 1.54$ A, $T_{on} = 2.18$ µs, and $T_{off} = 1$ µs in coded form. From both main effect plot and desirability function, it was established that lower I, higher T_{on}, and lower T_{off} is preferred for minimum *TWR.*

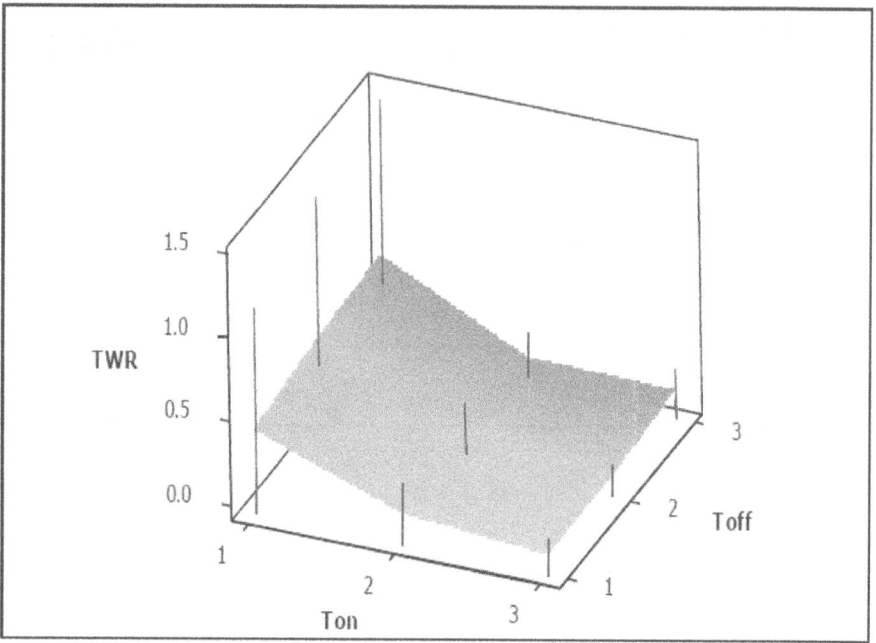

FIGURE 5.5   Surface plot for $T_{on} \times T_{off}$.

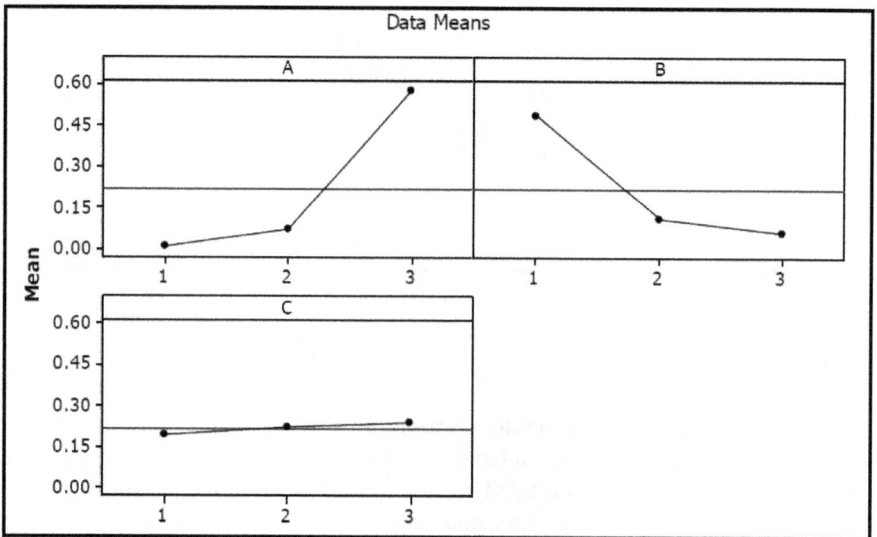

FIGURE 5.6   Main effect plot showing the effect of EDM factors on *TWR*.

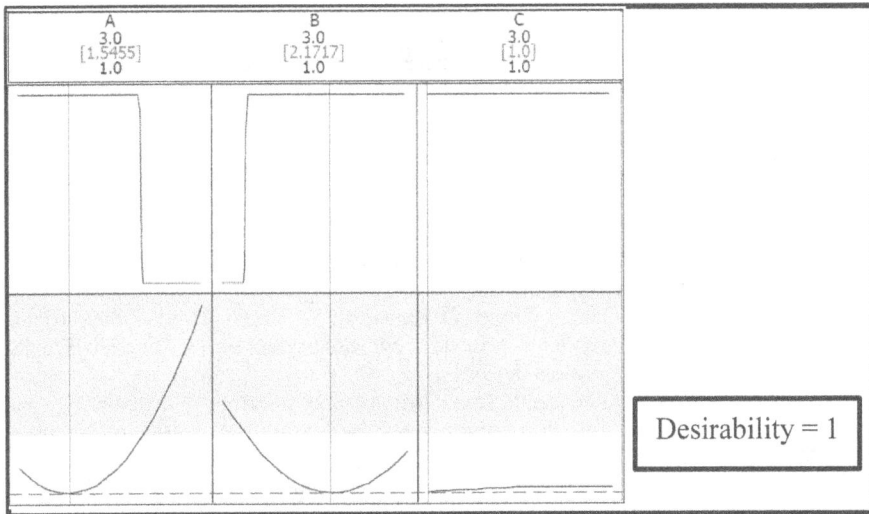

A 3.0 [1.5455] 1.0	B 3.0 [2.1717] 1.0	C 3.0 [1.0] 1.0

Desirability = 1

FIGURE 5.7 Optimized factor setting for minimal *TWR* using desirability function.

5.5 CONCLUSIONS

The present study investigates the wearing of the copper electrode (*TWR*) during the EDM of AISI 1035 steel. Electrode wear is estimated varying three important EDM factors, viz. current (*I*), spark on time (T_{on}), and spark off time (T_{off}). Results are analysed with the help of an ANOVA table, A-D plot, and 3D surface plots. After the determination of significant EDM factors affecting the *TWR*, the optimal set of factors which yield lower *TWR* is determined using the main effect plot and desirability function approach. Important conclusions drawn are:

1. Electrode wear in EDM is affected by a number of parameters including electrode material, electrode shape, carbon layer deposition on electrode, diameter of electrode, and machining factors.
2. While it is inevitable to compromise over other EDM parameters, it is more appropriate to choose an optimal range of EDM machining factors for minimization of *TWR*.
3. EDM of AISI 1035 using copper electrode demonstrates that *TWR* is significantly affected by current (*I*) and spark on time (T_{on}). At higher *I*, intensity of spark is higher, resulting in more *TWR*.
4. It was established that *TWR* is mostly affected by *I* at a low value of T_{on}. Higher T_{on} leads to decreases heat transfer to the electrode, and hence *TWR* decreases.
5. Spark off time (T_{off}) has minimal effect on *TWR*, especially at a higher value of T_{off}. At higher T_{off}, ample time is available for regaining of dielectric strength, which results in lower *TWR* in the next EDM cycle.
6. The main effect plot and optimization using the desirability method showed that minimal *TWR* is obtained at lower *I*, higher T_{on}, and lower T_{off}.

REFERENCES

1. N. M. Abbas, D. G. Solomon, M. F. Bahari, A review on current research trends in electrical discharge machining (EDM), International Journal of Machine Tools and Manufacture, 47 (7–8), 1214–1228, 2007.
2. B. Nahak and A. Gupta, A review on optimization of machining performances and recent developments in electro discharge machining, Manufacturing Reviews, 6, 1–22, 2019.
3. J. E. A. Qudeiri, A. Zaiout, A. I. Mourad, M. H. Abidi, A. Elkaseer, Principles and characteristics of different EDM processes in machining tool and die steels, Applied Sciences, 10 (6), 2082, 2020.
4. A. K. Singh, R. Mahajan, A. Tiwari, D. Kumar, R. K. Ghadai, Effect of dielectric on electrical discharge machining: A Review, IOP Conference Series: Materials Science and Engineering, 377, 012184, 2018.
5. P. Sadagopan, B. Mouliprasanth, Investigation on the influence of different types of dielectrics in electrical discharge machining, The International Journal of Advanced Manufacturing Technology, 92, 277–291, 2017.
6. C. Sanghani, D. Acharya, Effect of various dielectric fluids on performance of EDM: A review, Trends in Mechanical Engineering & Technology, 6, 55–71, 2016.
7. A. Equbal, M. I. Equbal, A. K. Sood, An investigation on the feasibility of fused deposition modelling process in EDM electrode manufacturing, CIRP Journal of Manufacturing Science and Technology, 26 10–25, 2019.
8. N. Mohri, M. Suzuki, M. Furuya, N. Saito, Electrode wear process in electrical discharge machining, Annals of the CIRP, 44(1), 165–168, 1995.
9. D. T Pham, A. Ivanov, S. Bigot, K. Popov, S. Dimov, A study of micro-electro discharge machining electrode wear, Proceedings of the Institution of Mechanical Engineers, Part C, Journal of Mechanical Engineering Science, 221, 605–612, 2007.
10. S. Kar and P. K. Patowari, Electrode wear phenomenon and its compensation in micro electrical discharge milling: A review, Materials and Manufacturing Processes, 33 (14), 1491–1517, 2018.
11. A. A. Khan, Electrode wear and material removal rate during EDM of aluminum and mild steel using copper and brass electrodes, The International Journal of Advanced Manufacturing Technology, 39, 482–487, 2008.
12. S. Laurenţiu, S. Hans-Peter, D. Oana, C. Margareta, G. Lorelei, G. Irina, Electrode tool wear at electrical discharge machining, Key Engineering Materials, 504–506, 1189–1194, 2012.
13. L. Slatineanu, O. Dodun, I. Carp, M. Coteata, I. Besliu, Tool electrode wear in electrical discharge of small diameter holes, MATEC Web of Conferences, 94, 03013, 1–7, 2017.
14. S. Heidari, A. Afsari, M. A. Ranaei, Increasing wear resistance of copper electrode in electrical discharge machining by using ultra-fine-grained structure, Transactions of the Indian Institute of Metals, 73, 2901–2910, 2020.
15. L. Hamdi, L. B. Toumi, Z. Salem, K. Allia, Full factorial experimental design applied to methylene blue adsorption onto Alfa stems, Desalination and Water Treatment, 57 (13), 6098–6105, 2015.
16. A. Equbal, A. K. Sood, M. A. Equbal, M. I. Equbal, An investigation on material removal rate of EDM process: A response surface methodology approach, World Academy of Science, Engineering and Technology International Journal of Mechanical and Mechatronics Engineering, 11 (4), 856–861, 2017.
17. A. Equbal, M. I. Equbal, A. K. Sood, PCA-based desirability method for dimensional improvement of part extruded by fused deposition modelling technology, Progress in Additive Manufacturing, 4, 269–280, 2019.

18. S. Arooj, M. Shah, S. Sadiq, S. H. I. Jaffery, S. Khushnood, Effect of current in the EDM machining of aluminum 6061 T6 and its effect on the surface morphology, Arabian Journal for Science and Engineering, 39(5), 4187–4189, 2014.
19. R. Kumar, O. P. Sahani, M. Vashista, Effect of EDM process parameters on tool wear, Journal of Basic and Applied Engineering Research, 1 (2), 53–56, 2014.
20. H. Singh, A. Singh, Effect of pulse on/pulse off on machining of steel using cryogenic treated copper electrode, International Journal of Engineering Research and Development, 5 (12), 29–34, 2013.

6 Achieving Optimal Efficiency in Manufacturing through Reinforced PA 3D Printed Parts Generated by FDM Technology

Aissa Ouballouch[1], Rachid El Alaiji[2],
Mohammed Sallaou[3], Aboubakr Bouayad[3],
Hamza Essoussi[3], Said Ettaqi[3], and Larbi Lasri[3]
[1]University of Hassan II, Morocco
[2]University of Abdelmalek Essaâdi, Morocco
[3]University of Moulay Ismail, Morocco

CONTENTS

6.1 Introduction 114
6.2 Experimental Procedure 116
    6.2.1 Materials, 3D Printer and Sample Preparation.................. 116
    6.2.2 Process Parameters and Experimental Set-up................... 117
        6.2.2.1 Surface Roughness... 119
        6.2.2.2 Dimensional Accuracy and Repeatability 119
        6.2.2.3 Tensile Tests and Total Costs 120
6.3 Results and Discussion 121
    6.3.1 Surface Roughness 121
        6.3.1.1 Impact of Print Speed 121
        6.3.1.2 Impact of Extrusion Temperature 122
        6.3.1.3 Impact of Layer Thickness 123
    6.3.2 Dimensional Accuracy and Repeatability 123
        6.3.2.1 Influence of Print Speed 123
        6.3.2.2 Influence of Extrusion Temperature 126
        6.3.2.3 Influence of Layer Thickness................................ 127

DOI: 10.1201/9780367822385-6

        6.3.3   Mechanical Properties Results 128
                6.3.3.1   Effect of Extrusion Temperature 130
                6.3.3.2   Effect of Print Speed.. 131
                6.3.3.3   Effect of Layer Thickness.. 132
        6.3.4   Build Time and Total Cost.. 132
                6.3.4.1   Build Time 135
                6.3.4.2   Total Cost.. 136
6.4   Conclusion 137
References... 138

6.1   INTRODUCTION

Over the years, the use of additive manufacturing (AM) techniques is gained importance in diverse applications domain [1–6]. In particular, fused deposition modeling (FDM) process is the most commonly utilized owing to its easy use, low cost, affordability of both materials and machines, etc. [7]. This technique is based on material extruding principle (Figure 6.1). In detail, the feedstock material in the form of a neat or reinforced polymer filament with a circular section (often a diameter of 1.75 mm) is introduced into a liquefier via drive rollers and heated to a semi-molten state using an electrical resistance, which enables it to pass through a moveable hot end nozzle. Then, the extruded line is deposited onto a building platform and solidified. With the moving of a hot end nozzle on an x-y axis system of 3D machine gantry, the deposition of adjacent lines is done and the entire layer is created. The building platform moves downward along the z-axis to deposit the subsequent layer. The cycle is repeated until the part is printed completely. However, the most used and compatible thermoplastics for FDM method are limited to those which have a low melting temperature, are much available and easy to process. Among these filaments, polylactic acid (PLA), acrylonitrile butadiene styrene (ABS). With these state of starting materials and working principle, the manufacture using FDM involves the setting of various PPs and the resulted parts depends on the combination of these PPs. Thus, the optimal selection of operating conditions has to be more evaluated and understood. Also, the majority of limitations which impair the functionality of the FDM end use parts results in the non-suitable combination of machine and fabrication settings.

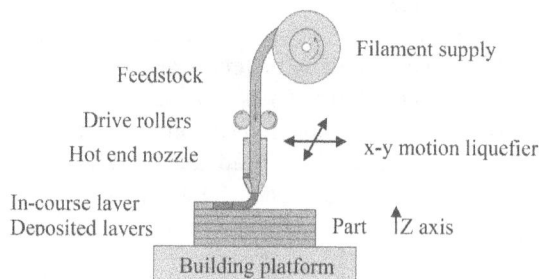

FIGURE 6.1   FDM process principle and schematic [8].

Dimensional inaccuracy, mediocre surface roughness, internal porosities formation and poor mechanical properties are the frequent drawbacks. The next lines present the efforts that have been undertaken to enhance the functionality of FDM components.

Interestingly, improving knowledge regarding the overall influence of the whole process is crucial to reach a desirable functionality and reproducibility. By examining the published literature, it can be noted that the attempts to overcome the above-mentioned limitations range between the development of raw materials, single and multiple ones [9–11], post-processing [12–21], reinforcement in particle and blend forms [22–31] and fibers (milled, chopped or continuous) [32–40] and investigation of PPs [41–76]. The latter is the most extensively studied as it is important regardless of adopted materials and methods. Basing on the review of literature, the main conclusions about optimization of processing parameters are the following: efficiency of study results is hampered by lack of standards for FDM technique and testing and limited or missing information about certain experiment elements. Therefore, an incongruence of test and operating settings was observed. For instance, with the smallest layer thicknesses (LT), the maximum performance is reached [77–78], while another work [51] showed that rising layer height leads to decreasing then increasing of tensile strength of tested specimens. In addition, a full and valid combination of raw material, 3D machine and all manufacturing settings that have not been considered in almost of these works contributed to the variability of findings. In other words, generalizing of the results and their comparison should be carried out carefully or should not be made. Another statement concerns the obvious effect and significance of some factors such as infill percent, build direction and raster orientation. These parameters, air gap and LT, have been reported as the key influencing operating conditions and have been actively assessed. But, the focus was on neat polymers such as ABS and PLA more than other materials. Another conclusion can be drawn; the tensile test is the most evaluated in comparison to other mechanical properties, surface roughness (Ra), quality part and total cost. Regarding the temperature in the FDM technique, few researches dealt with its effect taking into account that the process temperatures strongly affect the filament bonding and its viscosity. In this context, much researches dealing with the optimizing of PPs, especially those which were insufficiently studied, with variability in outcomes and for polymer composites, are necessary. Functionality in terms of mechanical performance, dimensional accuracy and quality surface should be the focus as this corresponds to real manufacturing conditions and applications.

Unlike previous works, in this study, a detailed investigation of performance and quality of 3D printed polyamide composites (with chopped glass fiber and Kevlar fiber reinforcement) was conducted. The comparison of these composites with fabricated additively neat polyamide and ABS parts and those processed by injection molding was carried out. And it involves mechanical performance (tensile test), quality (dimensional accuracy and surface roughness) and total cost. Full data on 3D printing conditions, tests and measurements were given. Thus, the overall influence of extrusion temperature (ET), LT and deposition speed was assessed. Interestingly, a specific comparison was done carefully, taking into account as possible raw material, 3D printer, reinforcement, performed test, testing equipment and other manufacture settings for better conclusions. This work is a completion of the published paper [52].

6.2 EXPERIMENTAL PROCEDURE

6.2.1 MATERIALS, 3D PRINTER AND SAMPLE PREPARATION

Chopped glass fiber reinforced polyamide (GRPA) and chopped Kevlar fiber reinforced polyamide (KRPA) filaments were used to evaluate the FDM process of nylon composites. They are in filament form with a diameter of 1.75 mm (TAGin3D as brand, corextrusion group as supplier). They are available and known commercially by TECHStrong and TECHArmed, respectively [53–54]. They have a protective thin skin to reduce the abrasive effect of reinforcement on the liquefier and also moisture impact. They consist of nylon polymer (neat PA) and 15% of glass fiber and 10% Kevlar fiber with 4 mm in length, respectively. For a comparison purpose, neat PA and ABS filaments were processed for one parameter selection (reference levels). Their brand and supplier are similar to those of polyamide composites. These neat filaments are provided in the market by their commercial name TECHLineTM and UNIVERSALAbsTM [55]. The main properties of used filaments are presented in Table 6.1.

Figure 6.2 shows a photograph of the Volumic STREAM 30 PRO MK2 3D machine, which is used to manufacture all the test specimens [61–62]. It is developed by the French manufacturer Volumic. This 3D printer uses various polymer materials such as PLA, ABS, NinjaFlex, and Nylon; its typical size nozzle is 0.4 mm. As all desktop FDM systems, STREAM 30 Pro MK2 can be controlled with any open-source

TABLE 6.1
Properties of Used Filaments

Properties	Standards	PA	ABS	PA + Glass Fiber (GRPA)	PA + Kevlar Fiber (KRPA)
Ultimate tensile strength [MPa]	ISO527 [56]	34.4	33.9	More than 175% of ABS strength	
Tensile modulus [MPa]	ISO 527	579.0 MPa	1681.5		
Strain at break [%]	ISO 527	20%	4.8%		
MRF [g/10 min]	ISO 1133 [57]	6.2 g/10 min	4.1		
Melting temperature [°C]	ISO 294 [58]/ ISO 11357 [59]	185–195°C	225–245°C		
Density	ISO 1183 [60]	1.14	1.10	1.37	1.18
Fiber diameter [mm]				0.215	0.215
Fiber length [mm]				4	4
Fiber content [%]				15%	10%
				Flexibility, resistance to impact, to fatigue, to chemical agents, to tensile, to wear and abrasion and to temperature, high resistance to weight ratio	
					Low moisture absorption, dimensional stability

FIGURE 6.2   3D printer used in the present research.

software. The geometry of additively manufactured samples was modeled using Catia V5 software exported as an STL file and imported to the 3D printing software Simplify 3D for G-code generation. Examined PA specimens were prepared in compliance with DIN EN ISO 527, having dimensions depicted in Figure 6.3.

6.2.2 PROCESS PARAMETERS AND EXPERIMENTAL SET-UP

As it is well known that PPs affect the performance and functionality of components processed by the FDM technique. The FDM settings investigated in this paper are tabulated in Table 6.2. A total of 78 reinforced PA samples were printed, where 13 samples run for each type of reinforcement and for each sample run 3 specimens were built using the same parameters setting to analyze the repeatability of FDM process (Table 6.3). In addition, three samples of neat ABS and those of neat PA were created considering their parameters reference levels to make a valid comparison.

For fixed parameters, they are outlined in Table 6.4. A range of evaluated printing conditions was selected based on literature study, serial trials/preliminary investigations and recommended intervals of material and 3D machine suppliers and slicing software. For instance, the strongest printing orientation and significant build direction are coincided with the pull direction during the deposition; flat orientation

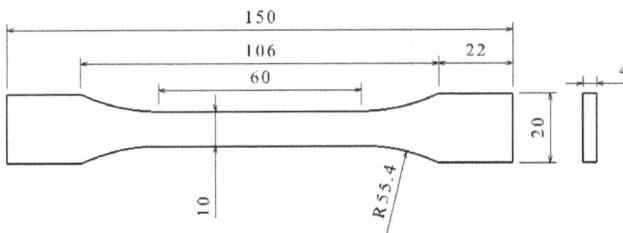

FIGURE 6.3   Nominal dimensions in mm according to the ISO 527 standards [63].

TABLE 6.2
Investigated Parameters and Their Levels

Parameter	Values for GRPA and KRPA	Reference Levels
Extrusion temperature [°C]	245, 255, 265	245
Print speed [mm/s]	30, 40, 50, 60, 70	50
Layer thickness [mm]	0.1, 0.15, 0.2, 0.25, 0.3	0.2

TABLE 6.3
Sample Processing Parameters Specification

Sample	Material	Extrusion Temperature [°C]	Print Speed [mm/s]	Layer Thickness [mm]
1		245	50	0.2
2		245	40	0.2
3		245	60	0.2
4		245	70	0.2
5		245	30	0.2
6		225	50	0.2
7	GRPA	235	50	0.2
8		255	50	0.2
9		265	50	0.2
10		245	50	0.1
11		245	50	0.15
12		245	50	0.25
13		245	50	0.3
1		245	50	0.2
2		245	40	0.2
3		245	60	0.2
4		245	70	0.2
5		245	30	0.2
6		225	50	0.2
7	KRPA	235	50	0.2
8		255	50	0.2
9		265	50	0.2
10		245	50	0.1
11		245	50	0.15
12		245	50	0.25
13		245	50	0.3
#	Neat PA	245	60	0.2
	Neat ABS	240	60	0.2

TABLE 6.4
Fixed Printing Conditions for Reinforced Materials GRPA and KRPA

Information/Process Parameters	GRPA	KRPA
Bed temperature [°C]	70	70
Envelope temperature [°C]	25	25
Contours number	3	3
Bottom layers number	3	3
Top layers number	3	3
Layer thickness [mm]	0.2 (as reference)	0.2 (as reference)
Building orientation	XYZ (Flat)	XYZ (Flat)
Raster angle [°]:crisscross	+45/−45	+45/−45
Width road [mm]	0.4	0.4
Air gap [mm]	0	0
Nozzle diameter [mm]	0.4	0.4
Infill degree [%]	100	100
Ventilation [%]	0	0
Bottom layers speed	40% × print speed	40% × print speed
Infill pattern	Diamond	Diamond
Filament diameter [mm]	1.75+/−0.01	1.75+/−0.01
Filament suppliers	TAGIN3D/CorExtrusion	TAGIN3D/CorExtrusion
Adhesive spray	GeckoTek	GeckoTek
Support material	NA	NA
Fiber content percent	15%	10%
Short fiber length	4 mm	4 mm
Filament color	White	Ivory
CAD software	Catia V5	Catia V5
Slicing software	Simplify 3D	Simplify 3D
File transfer format	STL	STL

(Z direction, which is perpendicular to the layer plane XY). This build direction is optimal as it yields high strength and enables variations of process parameters.

6.2.2.1 Surface Roughness

Specimens surface roughnesses (Ra [μm]) were averaged into a single value for each sample. It concerns top layer roughness. Measurements were conducted via RT-90G roughness tester parameters according to the following standards: ISO 4287:1997 [64] and ISO 12085:1998 [56].

6.2.2.2 Dimensional Accuracy and Repeatability

All 3D printed samples for both materials KRPA and GRPA were measured and compared to created 3D model. In this study, 13 measurements for each specimen were conducted; they included total length (L), width (W), width of narrow section (Wn) and thickness (T). Figure 6.4 depicts these dimensions in detail. The measuring is ensured by means of a micrometer. The values of each dimension were averaged.

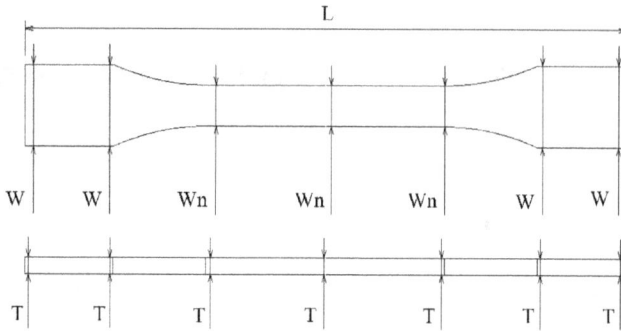

FIGURE 6.4   Locations of ISO 527 specimen dimensions.

6.2.2.3   Tensile Tests and Total Costs

A universal testing machine, Zwick/Roell Z050, equipped with a 50 kN load cell, was utilized to perform the tensile testing following the ISO 527 standards [56]. All specimens were tested at a speed of 10 mm/min at room temperature (~23°C). The load is increased until the specimen gets a break. The real time data recorded during the performing of the test were recovered by means of testing software and computer. The machine is shown in Figure 6.5.

The measuring of specimen's weight was ensured by means of a balance. As aforementioned, for each specimen number, three samples were prepared to obtain an average value of the measured properties and characteristics such as the ultimate tensile strength, build time and weight. In our case, total cost equation is mainly composed by the material cost per g, and fabrication cost per minute. These detailed components are illustrated in equation 6.1.

$$\text{Total cost}\,[\text{MAD}] = \left(\text{material}\,[\text{g}] \times \text{material cost per g}\,[\text{MAD}]\right)$$
$$+ \left(\text{built up time}\,[\text{min}] \times \text{machine cost}\,[\text{MAD}]\right) \tag{6.1}$$

As a detail, the cost of used glue is included in fabrication cost.

FIGURE 6.5   View of the tensile machine used in this study.

6.3 RESULTS AND DISCUSSION

6.3.1 SURFACE ROUGHNESS

The influence of PPs on surface roughness was assessed. Table 6.5 illustrates the values in terms of surface roughness (Ra). The impact of reinforcement nature on surface quality was clearly observed. Compared to pure materials, reinforced materials exhibit higher values of surface roughness.

6.3.1.1 Impact of Print Speed

The following graph highlights the impact of print speed (PS) on surface roughness for both reinforced polyamide materials. KRPA roughness is highly influenced by

TABLE 6.5
Measurements of Surface Roughness (Ra)

Specimen Number	Material	Extrusion Temperature [°C]	Print Speed [mm/s]	Layer Thickness [mm]	Surface Roughness Ra [µm]
1		245	50	0.2	5.041
2		245	40	0.2	6.427
3		245	60	0.2	8.262
4		245	70	0.2	5.787
5		245	30	0.2	7.258
6		225	50	0.2	8.013
7	GRPA	235	50	0.2	4.027
8		255	50	0.2	12.050
9		265	50	0.2	15.512
10		245	50	0.1	4.021
11		245	50	0.15	5.591
12		245	50	0.25	16.155
13		245	50	0.3	19.220
1		245	50	0.2	9.648
2		245	40	0.2	18.590
3		245	60	0.2	11.788
4		245	70	0.2	23.628
5		245	30	0.2	18.082
6		225	50	0.2	12.626
7	KRPA	235	50	0.2	14.623
8		255	50	0.2	12.122
9		265	50	0.2	21.777
10		245	50	0.1	5.411
11		245	50	0.15	14.049
12		245	50	0.25	29.281
13		245	50	0.3	21.040
	Neat PA	245	60	0.2	5.412
#	Neat ABS	240	60	0.2	5.424

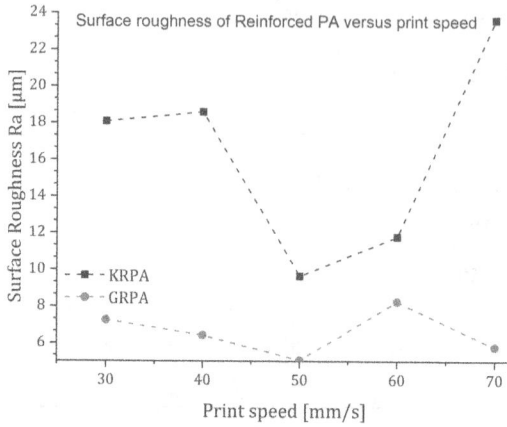

FIGURE 6.6 Surface roughness (Ra) versus print speed.

velocity compared to GRPA one. From Figure 6.6, it can be seen the significant impact of PS over the surface roughness of KRPA samples. Kevlar reinforcement tended to promote higher roughnesses when varying print velocity.

6.3.1.2 Impact of Extrusion Temperature

The graph below shows the surface roughness of reinforced PA specimens as a function of ET (Figure 6.7). The effect of ET due to the used filament was different for KRPA and GRPA samples. It is worth mentioning that the temperature has a direct relation with surface roughness. This trend can be interpreted by an unsuitable flow of extruded materials due to higher temperatures.

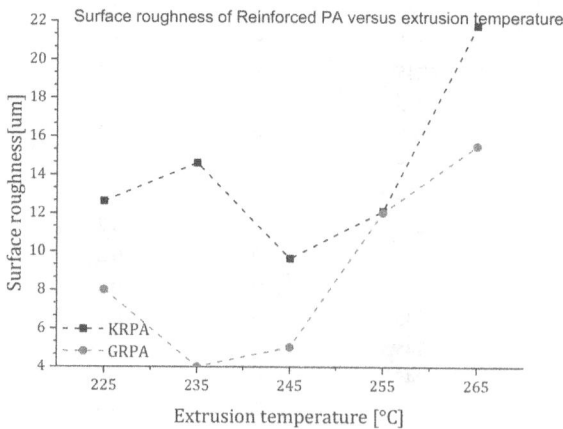

FIGURE 6.7 Surface roughness (Ra) versus extrusion temperature.

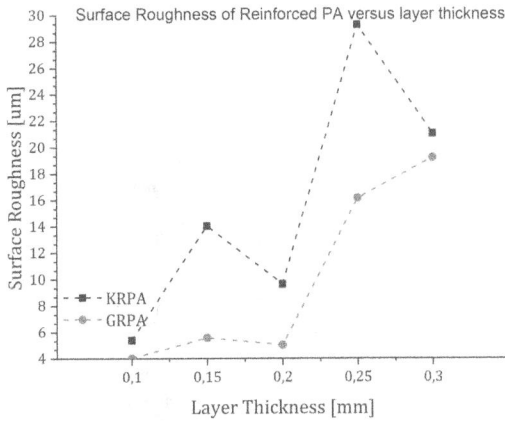

FIGURE 6.8 Surface roughness (Ra) versus layer thickness.

6.3.1.3 Impact of Layer Thickness

Figure 6.8 shows the variation of surface roughness of reinforced PA parts with the altering of LT. For both materials, there is an increase of roughness with rising of thickness. However, the impact in the case of glass reinforcement is lower than Kevlar reinforcement case. The impact of LT on surface quality can be explained by that a higher layer yields voids between deposited lines along the deposition direction. Thereby higher values of surface roughness (Ra).

6.3.2 DIMENSIONAL ACCURACY AND REPEATABILITY

The effect of process parameters on dimensional accuracy was observed. The observations in terms of a reduction in length and an increase in thickness have been confirmed in previous works [49, 79]. Also, an increment in widths was seen for all measured specimens. Results are presented in Table 6.6. Likewise, larger deviations for the dimensions variations of specimens under certain fabrication settings are shown in this table.

The dimensional error is calculated using the following equation:

$$\text{Error} = \text{Measured value} - \text{3D model value} \tag{6.2}$$

6.3.2.1 Influence of Print Speed

The following graph highlights the impact of PS on dimensional accuracy for both reinforced polyamide materials. Width and LT errors were negative, which means there is a reduction of these dimensions, while errors of length were positive with slight fluctuations. The effect of PS on width and thickness errors due to the used filament was different for KRPA and GRPA samples. The trend was not the same. A detailed description of these graphical data is presented below.

TABLE 6.6

Samples Measurements Averaged Results of Dimensions in mm

Specimen Number	Materials	Extrusion Temperature [°C]	Print Speed [mm/s]	Layer Thickness [mm]	Total Length [mm]	Width [mm]	Reduced Section Width [mm]	Thickness [mm]
3D model	–	–	–	–	150	20	10	4
1		245	50	0.2	149.82 (0.007)	20.35 (0.088)	10.428 (0.017)	4.056 (0.162)
2		245	40	0.2	149.32 (0.042)	20.336 (0.098)	10.423 (0.026)	4.09 (0.115)
3		245	60	0.2	149.345 (0.007)	20.263 (0.096)	10.341 (0.027)	4.11 (0.071)
4		245	70	0.2	149.505 (0.120)	20.253 (0.075)	10.305 (0.044)	4.078 (0.171)
5		245	30	0.2	149.505 (0.120)	20.296 (0.109)	10.36 (0.027)	4.13 (0.157)
6		225	50	0.2	149.525 (0.035)	20.262 (0.041)	10.363 (0.039)	4.13 (0.088)
7	GRPA	235	50	0.2	149.095 (0.021)	20.266 (0.095)	10.34 (0.059)	4.1 (0.126)
8		255	50	0.2	149.37 (0.014)	20.35 (0.067)	10.43 (0.031)	4.16 (0.146)
9		265	50	0.2	149.32 (0.084)	20.39 (0.075)	10.53 (0.019)	4.189 (0.099)
10		245	50	0.1	149.755 (0.007)	19.975 (0.049)	10.1 (0.035)	4.128 (0.124)
11		245	50	0.15	149.75 (0.014)	20.126 (0.070)	10.245 (0.022)	4.25 (0.143)
12		245	50	0.25	149.73 (0.028)	20.52 (0.073)	10.588 (0.042)	4.347 (0.090)
13		245	50	0.3	149.72 (0.014)	20.573 (0.087)	10.69 (0.038)	4.35 (0.094)
1		245	50	0.2	149.995 (0.035)	20.537 (0.032)	10.55 (0.039)	4.167 (0.130)
2		245	40	0.2	149.95 (0.099)	20.56 (0.096)	10.52 (0.026)	4.14 (0.148)
3		245	60	0.2	149.925 (0.007)	20.578 (0.062)	10.668 (0.035)	4.166 (0.177)
4		245	70	0.2	149.975 (0.021)	20.526 (0.097)	10.543 (0.089)	4.147 (0.119)
5		245	30	0.2	149.995 (0.021)	20.641 (0.056)	10.666 (0.017)	4.255 (0.117)
6		225	50	0.2	149.835 (0.177)	20.447 (0.030)	10.485 (0.130)	4.131 (0.092)
7	KRPA	235	50	0.2	149.86 (0.028)	20.481 (0.077)	10.51 (0.035)	4.199 (0.098)
8		255	50	0.2	149.83 (0.014)	20.608 (0.053)	10.678 (0.084)	4.243 (0.141)
9		265	50	0.2	149.83 (0.015)	20.672 (0.015)	10.70 (0.049)	4.215 (0.090)
10		245	50	0.1	149.78 (0.254)	20.158 (0.039)	10.226 (0.029)	4.09 (0.091)

(Continued)

TABLE 6.6 *(Continued)*

Samples Measurements Averaged Results of Dimensions in mm

Specimen Number	Materials	Extrusion Temperature [°C]	Print Speed [mm/s]	Layer Thickness [mm]	Total Length [mm]	Width [mm]	Reduced Section Width [mm]	Thickness [mm]
11		245	50	0.15	149.765	20.445	10.436	4.268
					(0.007)	(0.023)	(0.026)	(0.092)
12		245	50	0.25	149.865	20.71	10.603	4.515
					(0.007)	(0.079)	(0.207)	(0.089)
13		245	50	0.3	150.035	20.727	10.6	4.514
					(0.007)	(0.135)	(0.253)	(0.064)
#	Neat ABS	240	60	0.2	149.339	20.099	10.008	3.989
					(0.023)	(0.027)	(0.025)	(0.070)
	Neat PA	245	60	0.2	149.018	20.359	10.246	4.115
					(0.058)	(0.101)	(0.040)	(0.043)

Note: Standard deviation is depicted in brackets.

Dimensional error of KRPA specimens against PS is shown in Figure 6.9. From the graph, it can be seen that width and thickness errors remain approximately constant when the PS increases, except for the lower value 30 mm/s. At this print velocity, the errors are at their highest values 0.67 mm, 0.66 mm and 0.38 mm for width, width narrow and thickness, respectively. There is an explanation for this trend, a certain lower speed may lead to an over deposition of the extruded material at the same area, which causes dimensional inaccuracy in terms of bigger dimensions than those of CAD files. In addition, as well known, the ET and the PS are dependent; at a limit value of speed (minimum or maximum) for a given temperature, a special heat transfer or thermal behavior may be developed and affect the process in terms of dimensional error. Taking into account this statement, a couple of higher values of temperature and

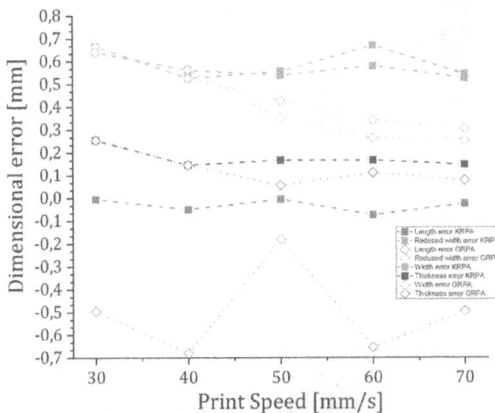

FIGURE 6.9 Dimensional error versus print speed.

speed should be avoided to guarantee the feasibility of the process and obtaining good parts. For length error, it is constant with negligible fluctuations, in contrast to other errors; the lowest value 0.000 mm is reached at the lowest PS 30 mm/s.

The same figure shows the variation of dimensional error of GRPA specimens with an increment of PS from 30 mm/s to 70 mm/s. The downward trend is observed for widths and thickness errors. In contrast, the length error is approximately constant with fluctuations at speeds 40 mm/s and 60 mm/s, where the errors are higher −0.6 mm and 0.5 mm, respectively. Regardless of the dimension, the lower errors are obtained at a maximum limit speed 70 mm/s. This finding can be explained by considering that with increased PS, the deposition for a given parameters combination will be suitable, and, therefore, the extruded portion line is deposited without much swelling. The fiber content (15%) in comparison to that of KRPA filament (10%) can be among the reasons of trend's difference between the two these materials under the same parameter values.

6.3.2.2 Influence of Extrusion Temperature

The graph below shows the dimensional error of reinforced PAs specimens as function of ET. The effect of ET on dimensional accuracy due to the used filament was different for KRPA and GRPA samples. Thus, the widths errors had an upward trend for KRPA while they remain constant for GRPA. For length error, it is approximately constant for KRPA while it has a downward trend for GRPA. It is worth mentioning that the temperature has a direct relation with shrinkage's effect [67]. The graphical representation of these variations is described in detail as follows.

In Figure 6.10, we can observe the significant impact of ET over the widths errors of KRPA samples. Higher ET tended to promote higher widths errors. These results were in agreement with previous findings [49, 68]. In this latter, the authors stated that a polymer is exhibited to be relaxed or expanded with an increased temperature. For thickness error, its variation was of slight significance. Nevertheless, a reduction of

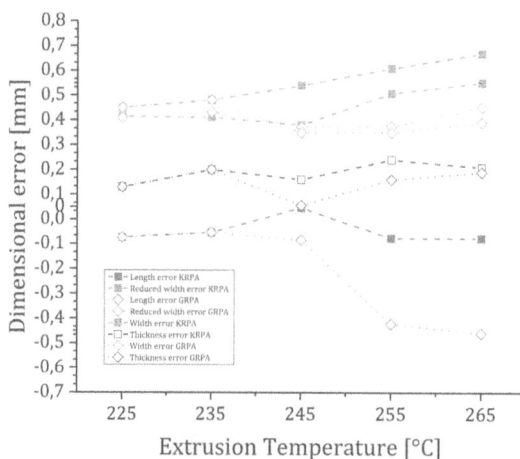

FIGURE 6.10 Dimensional error versus extrusion temperature.

error was achieved at the lowest value of temperature 225°C. Concerning the length error, it decreases as ET increases until 245°C, then it increases. The notable effect over the length error is observed when the temperature varies from 235°C to 255°C. This trend is in contrast with the drawn conclusion in [69]. Nonetheless, this may be due to a developed special temperature gradient and consequently, distortion manifested in a positive way by the reduced error of length. Looking at the same figure, it can be observed that the errors in all the measured dimensions were roughly constant except that of length, which experienced an upward trend. In other words, ET affected significantly only the length on GRPA specimens. This influence has been reported in [49], but the variations had not the same pattern. Nevertheless, the lowest errors were achieved at the lowest value of temperature 225°C for the length while for other dimensions at 245°C. This reduction of errors was not supported by [49].

6.3.2.3 Influence of Layer Thickness

The following graph highlights the impact of LT on the dimensional accuracy for both reinforced polyamide materials. The widths and LT errors were negative, which means there is a reduction of these dimensions, while errors of length were almost positive with some fluctuations. In general, the variation of dimensional errors is greatly affected by LT. This finding does not match the outcomes of a previous study [65]. A detailed description of these graphical data is presented below:

In Figure 6.11, we can observe clearly the significant effect of LT on dimensional accuracy, in particular for thickness and widths. The smaller errors were obtained with the lowest layer height 0.1 mm except that of overall length, whose lower error is reached with the slice thickness 0.2 mm. It may be interpreted by the fact that with this thickness and the values of other PPs, the shrinkage phenomenon becomes reduced yields the part to keep its length. In addition, the graph shows the slight effect over length error compared to those of other specimen dimensions, as stated in [49]. The thickness specimen equals 4mm, which is an integer multiple of 0.1 mm,

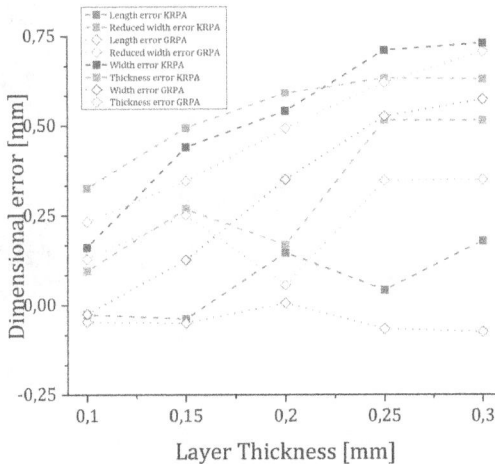

FIGURE 6.11 Dimensional error versus layer thickness.

0.2 mm and 0.25 mm. So, this explains why error values of thickness increase although the slice height is small. For instance, at 0.15 mm, the error value is 0.268 mm, while at 0.2 mm, it is worth 0.167 mm. For widths errors, they experienced an upward trend. These results are partially supported by the previous work [49].

In Figure 6.11, widths errors rise as LT increases. An increment in LT from 0.1 mm to 0.3 mm, corresponded to an increment in terms of width error and reduced width error from 0 mm to 0.4 mm and from 0.1 mm to 0.6 mm, respectively. In the case of thickness error, it had an upward trend as function of LT with some fluctuations due to the values of slice thickness in relationship with the specimen thickness, as explained before. An other reason for these fluctuations resides in the combination parameters effect. Concerning the length error, it decreases and then increases as the layer height increases. This may be due to the previous above-mentioned. Additionally, the deformation of printed specimens during their processing (shrinkage, distortion) can yield reduction and variation of the length for a given FDM parameters combination. For example, as observed, when the widths and the thickness are measured in some areas of the part, the value is lower and in other location is higher, which may influence the volume and the length of the specimen.

6.3.3 MECHANICAL PROPERTIES RESULTS

Table 6.7 resumes the mechanical test values and total cost. The tensile test results for each of the reinforced Nylon polymers were compared with data obtained from the test of neat Nylon and ABS. In addition to other common fabrication conditions,

TABLE 6.7
Tensile Test Results (UTS)

Specimen Number	Material	Extrusion Temperature [°C]	Print Speed [mm/s]	Layer Thickness [mm]	UTS [MPa]
1		245	50	0.2	24.52 (0.579)
2		245	40	0.2	23.46 (0.739)
3		245	60	0.2	23.67 (0.191)
4		245	70	0.2	21.97 (0.319)
5		245	30	0.2	26.38 (0.070)
6	GRPA	225	50	0.2	23.05 (1.735)
7		235	50	0.2	22.60 (0.723)
8		255	50	0.2	23.95 (0.106)

(Continued)

TABLE 6.7 *(Continued)*
Tensile Test Results (UTS)

Specimen Number	Material	Extrusion Temperature [°C]	Print Speed [mm/s]	Layer Thickness [mm]	UTS [MPa]
9		265	50	0.2	26.29 (0.127)
10		245	50	0.1	21.22 (0.010)
11		245	50	0.15	23.55 (0.049)
12		245	50	0.25	28.95 (0.350)
13		245	50	0.3	29.50 (1.042)
1		245	50	0.2	35.20 (0.070)
2		245	40	0.2	35.27 (0.576)
3		245	60	0.2	35.34 (0.629)
4		245	70	0.2	34.99 (0.637)
5		245	30	0.2	34.66 (0.116)
6		225	50	0.2	34.66 (0.533)
7	KRPA	235	50	0.2	35.37 (0.718)
8		255	50	0.2	36.48 (1.286)
9		265	50	0.2	37.41 (1.145)
10		245	50	0.1	25.64 (0.346)
11		245	50	0.15	39.55 (0.077)
12		245	50	0.25	35.01 (0.051)
13		245	50	0.3	31.24 (3.853)
#	Neat PA	245	60	0.2	28.30 (3.101)
	Neat ABS	240	60	0.2	28.50 (3.177)

Note: Standard deviation is depicted in parentheses.

the same 3D printer was used to fabricate all these materials that were supplied by the same company for a better and valid comparison. As to used feedstock filaments, GRPA is stronger than ABS, about 75%, as mentioned in the supplier data sheet. However, our experimental results showed that there is a slight improvement in GRPA strength compared to that of ABS. With regard to pure and reinforced PA comparison, Kevlar yields a 30% of enhancement in tensile strength. Furthermore, it is worth noting that a higher standard deviation was observed in the results, with LT = 0.3 mm and ET = 225°C for GRPA specimens. With regard to KRPA samples, larger deviations were observed in the findings with LT = 0.3 mm and ET = {255°C, 265°C}. Neat materials PA and ABS exhibit higher standard deviations more than used reinforced materials. This observation can be explained by considering the effect of reinforcement in enhancing printing process stability.

6.3.3.1 Effect of Extrusion Temperature

Figure 6.12 offers a graphical representation of data on ET influence on tensile properties. Ultimate tensile strength increased with the rising of ET (with an increment of 10 mm/s) for both filaments. This is obvious as the fusion interlines and interlayers bonding and density are improved. These findings support previous studies [19, 49, 70], while it was reported in [69] that the optimal temperature is reached at the lowest temperature value of the adopted range 210°C–230°C. It is interesting to note that after a certain high value of temperature, the performance takes a downward trend. In regard to GRPA composite, it experienced slight fluctuations (about 3°C and 2.5°C of difference) at temperatures 235°C and 255°C, which is in accordance with the results obtained in [72], where the tensile strength increases then decreases when the temperature varies from 218.5°C to 241.5°C. The same conclusion was drawn in [40, 73]. According to experimental observations, in

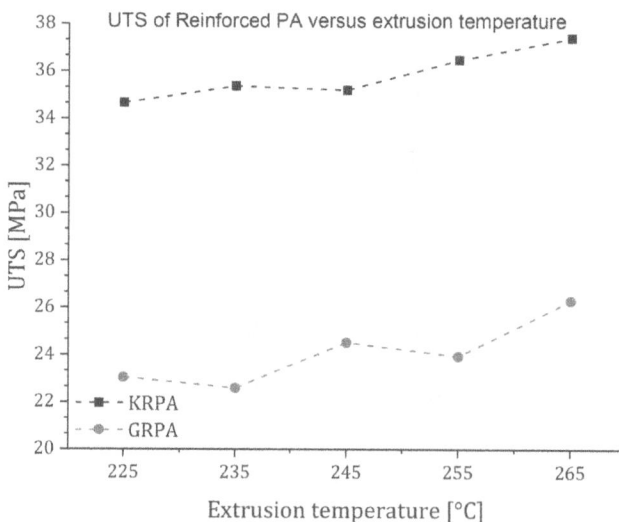

FIGURE 6.12 UTS versus extrusion temperature.

our case, this can be explained by changing of environment conditions and a delay ET to achieve the nozzle. It may be due to a special heat transfer or temperature gradient developed at this temperature. Comparing the two reinforcements, tensile strength is in the order: Kevlar fiber > glass fiber. Overall, the temperature did not highly affect tensile strength. And the gain in terms of performance is 12% for GRPA and 7% for KRPA. In other words, the highest strength values are reached at a limit temperature 265°C.

6.3.3.2   Effect of Print Speed

The graph in Figure 6.13 illustrates the evolution of UTS of nylon composites in function of PS ranging between 30 mm/s and 70 mm/s with an increment of 10 mm/s. Strength nylon composite KRPA remains steady while GRPA has a downward trend as PS rises. However, there is an exception at a PS 50 mm/s where strength increases with 1 MPa in comparison to this of 40 mm/s. Thus, mechanical properties are not highly affected by the PS, which confirms results reported in literature, respectively tensile strength and flexural strength [49, 74, 75]. PS contributes in this manner; it should be higher enough to have an appropriate deposition but not much for suitable impregnation period, pressure and better overlapping between adjacent deposited lines. However, it depends on the ET as they cannot be both at their high values. Considering the type of material, the trend of both filaments is different, which may be due to the fact that the PS range for nylon reinforced by short Kevlar fibers lead to a reduction in impregnation period and bonding degree as well as increasing in porosities. This explanation is supported by [76]. In addition, it was deducted in [77] that effect of print velocity over ultimate tensile strength depends on other parameters fixed such as LT. The strengths of KRPA keep higher than those of GRPA.

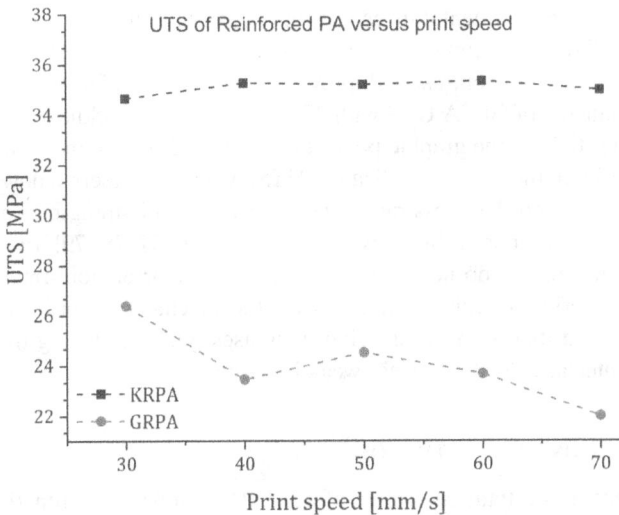

FIGURE 6.13   UTS versus print speed.

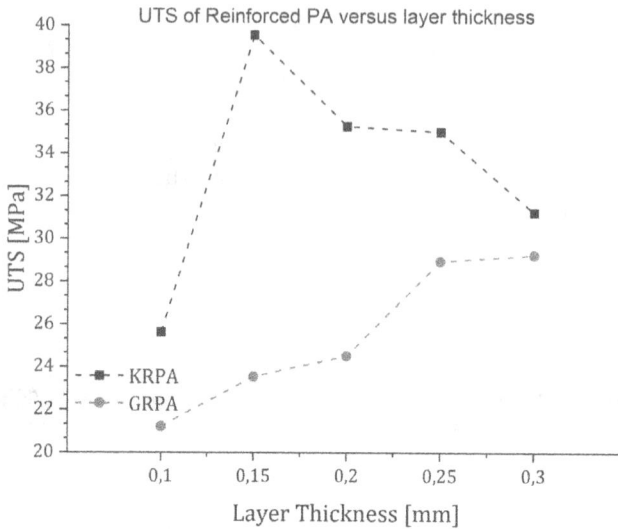

FIGURE 6.14   UTS versus layer thickness.

6.3.3.3   Effect of Layer Thickness

Figure 6.14 shows UTS as a function of LT, among other process parameters. The effect of LT on mechanical properties due to raw material was different for KRPA and GRPA samples. As to GRPA, higher LT tended to promote higher tensile strength. This can be interpreted by that part with less layers is stronger as the number of interfaces is lower and distortion effects due to heating and cooling cycles are minimized with an increase of strength [74, 77, 78]. Material composition and PPs combination also contributed to this enhancement. These results support reported published studies [37, 49, 78]. In experimental investigation [37], tensile strength increases with the evolution of LT from 0.2 mm to 0.3 mm then declines at 0.4 mm. On the other hand, these outcomes were in contrast with previous works [49–51, 76, 80–82]. The graph indicates enhancing of GRPA UTS with 27% from 0.1 mm thickness to 0.3 mm.

Concerning KRPA, the graphic presentation showed that an increase of LT from 0.1 mm to 0.15 mm increased significantly UTS, which then keeps a downward trend until the limit of used thickness range. This correlation of strength as a function of slice thickness was reported in many published works [37, 78, 79]. In other studies [49], drawn conclusions do not support this observed correlation. Interpretation of contrast in the effect of this parameter over the mechanical performance can be made considering that when layer height increases, yields reducing the number of layers, thus interfaces leading to less weak bonding.

6.3.4   BUILD TIME AND TOTAL COST

Table 6.8 depicts the total cost of reinforced PA samples as a function of investigated PPs. Total cost in terms of printing time is directly related to LT and PS. Thereby, total cost decreases as LT increases and increasing of PS leads to their

TABLE 6.8
Build Time and Total Cost Versus PPs. Standard Deviation Is Depicted in Brackets

Specimen Number	Material	Extrusion Temperature [°C]	Print Speed [mm/s]	Layer Thickness [mm]	Build Time [min]	Total Cost [MAD]
1		245	50	0.2	45	10.82 (0.017)
2		245	40	0.2	51	10.95 (0.010)
3		245	60	0.2	40	10.74 (0.002)
4		245	70	0.2	37	10.66 (0.007)
5		245	30	0.2	64	11.20 (0.031)
6		225	50	0.2	40	10.77 (0.007)
7	GRPA	235	50	0.2	45	10.82 (0.027)
8		255	50	0.2	45	10.85 (0.004)
9		265	50	0.2	45	10.87 (0.001)
10		245	50	0.1	89	11.53 (0.008)
11		245	50	0.15	60	11.30 (0.005)
12		245	50	0.25	36	11.10 (0.001)
13		245	50	0.3	29	10.68 (0.005)
1		245	50	0.2	45	8.83 (0.012)
2		245	40	0.2	51	8.88 (0.027)
3		245	60	0.2	40	8.73 (0.015)
4		245	70	0.2	37	8.64 (0.027)
5		245	30	0.2	64	9.22 (0.071)
6		225	50	0.2	40	8.70 (0.020)
7		235	50	0.2	45	8.90 (0.077)
8	KRPA	255	50	0.2	45	8.91 (0.003)

(Continued)

TABLE 6.8 *(Continued)*
Build Time and Total Cost Versus PPs. Standard Deviation is Depicted in Brackets

Specimen Number	Material	Extrusion Temperature [°C]	Print Speed [mm/s]	Layer Thickness [mm]	Build Time [min]	Total Cost [MAD]
9		265	50	0.2	45	8.95 (0.008)
10		245	50	0.1	89	9.65 (0.001)
11		245	50	0.15	60	9.41 (0.006)
12		245	50	0.25	36	9.16 (0.006)
13		245	50	0.3	29	8.83 (0.024)
#	Neat PA	245	60	0.2	113	7.2 (0.021)
	Neat ABS	240	60	0.2	113	3.56 (0.110)

decrease. Additionally, rising of ET heighten total cost. However, the significance of ET over total cost is lower than that of two other evaluated parameters. These tabulated results are illustrated in Figures 6.15–6.19 for detailed discussion. In addition, standard deviations are shown Table 6.8. The majority of deviations for total cost values under certain printing conditions are smaller. A higher standard deviation was observed in the results of the neat ABS specimen.

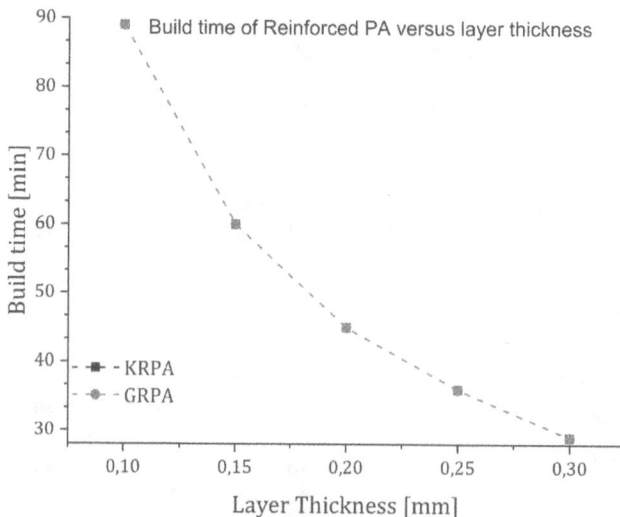

FIGURE 6.15 Build time versus layer thickness.

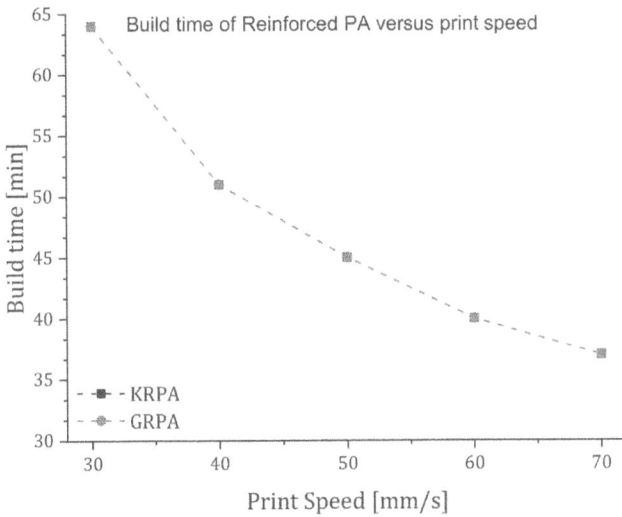

FIGURE 6.16   Build time versus print speed.

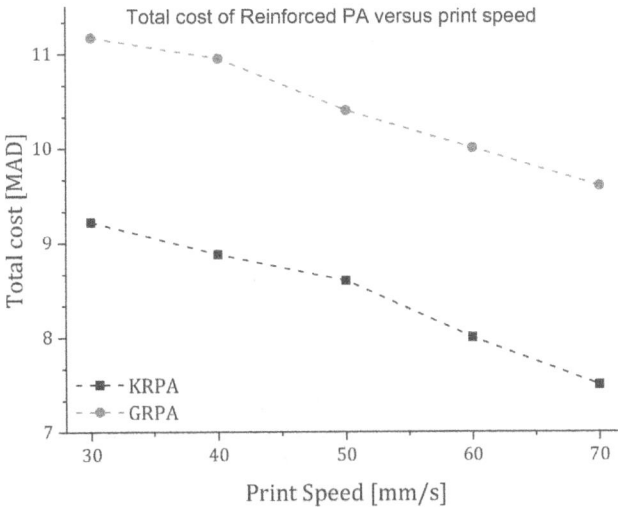

FIGURE 6.17   Total cost of reinforced PA specimens versus print speed.

The results in average values of total cost including build time of reinforced PA samples with different levels of process parameters are shown in Table 6.8.

6.3.4.1   Build Time

Figures 6.15 and 6.16 depict a significant drop in build time with increased PS and LT. For ET variations, build time remains constant.

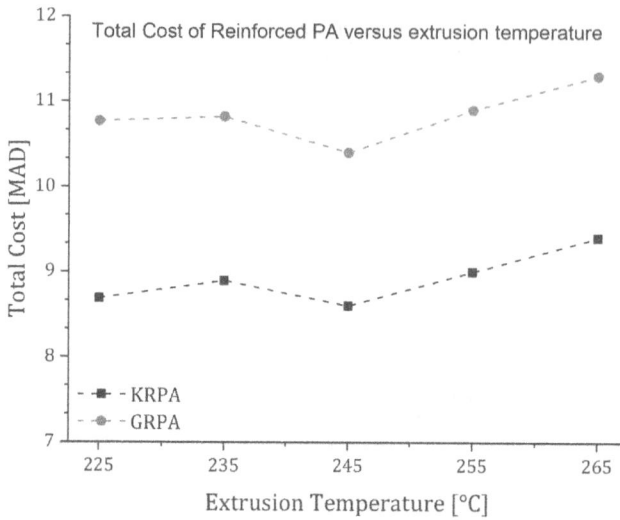

FIGURE 6.18 Total cost of reinforced PA specimens versus extrusion temperature.

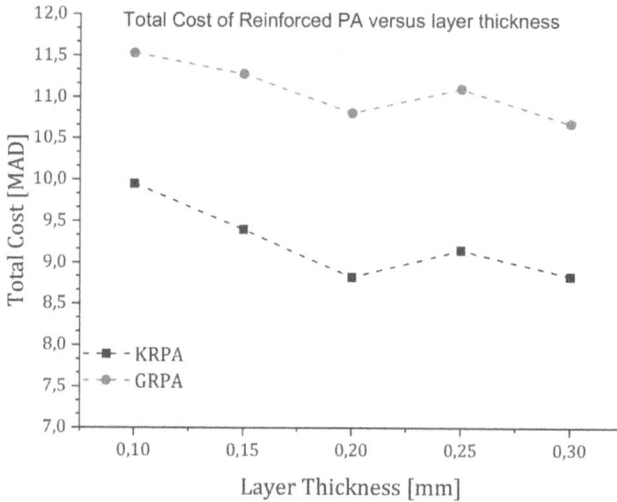

FIGURE 6.19 Total cost of reinforced PA specimens versus layer thickness.

6.3.4.2 Total Cost

Figures 6.17–6.19 depict a significant drop in total cost with increased PS and LT, contrary to ET, which yields a slight rise in total cost when it increases. This difference can be explained considering that increasing of weight samples (hence material price) with an increased temperature is less significant than that of rising of printing time due to higher values of slice thickness and print velocity. Thus, LT showed the highest reduction for the total cost (Figure 6.19), while ET resulted in the lowest one (Figure 6.18) regardless of printed samples (GRPA or KRPA). From a material point of view, all the figures show that the total cost of reinforced glass polyamide had the highest values, which are mainly associated to the higher price of this filament. For the effect of LT, there is a particular case, LT = 0.2 mm, where total costs (10.82 MAD and 8.83 MAD) are lower than those of LT = 0.25 mm (11.10 MAD and 9.16 MAD), respectively for GRPA and KRPA filaments. This exception is interpreted that at this layer height, the corresponding specimens' weights are lower than those of LT = 0.25 mm, affecting total cost. In terms of percent, an increment in the PS from 30 to 70 mm/s corresponded to a reduction of 5% in overall cost for GRPA, while for KRPA material, it is about 6.3%. In addition, rising LT from 0.1 to 0.3 mm caused a reduction of 7.4% and 8.5% in total costs of GRPA and KRPA parts, respectively. Moreover, regarding ET, its decrease from 265 to 225°C leads to a diminution of 1% and 2.8% in total cost of GRPA and KRPA samples, respectively. Thus, the impact of assessed settings from the cost point of view is clearly shown, and polyamide reinforced by Kevlar chopped fibers is more affected between two reinforced PAs.

6.4 CONCLUSION

Three controlling process parameters with different ranges were considered. Print speed (PS = {30, 40, 50, 60, 70} mm/s), layer thickness (LT = {0.1, 0.15, 0.2, 0.25, 0.3} mm) and ET (ET = {225, 235, 245, 255, 265} °C), were varied to obtain and investigate reinforced PA; KRPA and GRPA. For each parameter, a reference level is fixed and other parameters altered within the considered range in order to print samples. Tensile mechanical properties observed to be affected by ET, LT and reinforcement less than PS. Also, the layer effect is related to reinforcement content, and to raise tensile resistance, a higher ET is needed. Optimal parameters settings were related to reinforcement content. KRPA tensile strength was 39.55 MPa, while that of GRPA equaled 29.5 MPa. These values are more than that of neat PA. Besides, variations of tensile properties, related plots and experimental runs showed some benefits such as process stability, repeatability and extending parameters range, which is suitable for industrial applications. However, other aspects are to be considered, namely dimensional accuracy, surface roughness, lead time and total cost. Thereby, the dimensional accuracy of both PAs was assessed and found to be influenced by ET and LT more than PS and reinforcement. KRPA surface roughness was largely affected by process parameters more than GRPA. Total cost was found to be notably influenced by PS, LT and nature or reinforcement. Overall, conducted experimental works identified the trends of printed samples characteristics under key FDM parameters and their combination. In these, detailed data are provided, which allows their exploitation, unlike most previous studies, and helps understand results variability. They showed the ability of the FDM process to compete with conventional manufacturing processes in various industrial settings. The future direction to build on this work is to find a link between LT and percent content of reinforcement, considering other characteristics such as surface roughness and dimensional accuracy. Moreover, using finite element analysis can help to achieve promising results, in particular, coupled to an experimental approach.

REFERENCES

1. E. Kroll and D. Artzi, "Enhancing aerospace engineering students' learning with 3D printing wind-tunnel models," *Rapid Prototyp. J.*, vol. 17, no. 5, pp. 393–402, 2011.
2. P. Parandoush and D. Lin, "A review on additive manufacturing of polymer-fiber composites," *Compos Struct.*, vol. 182. Elsevier Ltd, pp. 36–53, December 15, 2017.
3. K. V. Wong and A. Hernandez, "A review of additive manufacturing," *ISRN Mech. Eng.*, vol. 2012, pp. 1–10, 2012.
4. S. V. Murphy and A. Atala, "3D bioprinting of tissues and organs," *Nat. Biotechnol.*, vol. 32, no. 8. Nature Publishing Group, pp. 773–785, 2014.
5. D. B. Short, "Use of 3D printing by museums: Educational exhibits, artifact education, and artifact restoration," *3D Print. Addit. Manuf.*, vol. 2, no. 4, pp. 209–215, January 2016.
6. X. Wang, M. Jiang, Z. Zhou, J. Gou, and D. Hui, "3D printing of polymer matrix composites: A review and prospective," *Composites Part B: Eng.*, vol. 110. Elsevier Ltd, pp. 442–458, February 01, 2017.
7. C. K. Chua, K. F. Leong, and C. S. Lim, *Rapid prototyping: Principles and applications, third edition*. World Scientific Pub Co Inc, 2010.
8. A. Ouballouch, R. Elalaiji, I. Ouahmane, L. Lasri, and M. Sallaou, "Finite element analysis of a FDM 3D printer liquefier," *Key Engineering Materials*, vol. 820. Trans Tech Publications Ltd, pp. 173–178, 2019.
9. F. Tamburrino, S. Graziosi, and M. Bordegoni, "The influence of slicing parameters on the multi-material adhesion mechanisms of FDM printed parts: An exploratory study," *Virtual Phys. Prototyp.*, vol. 14, no. 4, pp. 316–332, 2019.
10. H. Kim, E. Park, S. Kim, B. Park, N. Kim, and S. Lee, "Experimental study on mechanical properties of single- and dual-material 3D printed products," *Procedia Manuf.*, vol. 10, pp. 887–897, 2017.
11. J. F. Rodríguez, J. P. Thomas, and J. E. Renaud, "Mechanical behavior of acrylonitrile butadiene styrene (ABS) fused deposition materials. Experimental investigation," *Rapid Prototyp. J.*, vol. 7 , no. 3, pp. 148–158, 2001.
12. O. Kerbrat, P. Mognol, and J. Y. Hascoët, "A new DFM approach to combine machining and additive manufacturing," *Comput. Ind.*, vol. 62, no. 7, pp. 684–692, 2011.
13. P. M. Pandey, N. V. Reddy, and S. G. Dhande, "Improvement of surface finish by staircase machining in fused deposition modeling," *J. Mater. Process. Technol.*, vol. 132, no. 1, pp. 323–331, 2003.
14. V. Tiwary, P. Arunkumar, A. S. Deshpande, and V. Khorate, "Studying the effect of chemical treatment and fused deposition modelling process parameters on surface roughness to make acrylonitrile butadiene styrene patterns for investment casting process," *Int. J. Rapid Manuf.*, vol. 5, no. 3–4, pp. 276–288, 2016.
15. A. Garg, A. Bhattacharya, and A. Batish, "Chemical vapor treatment of ABS parts built by FDM: Analysis of surface finish and mechanical strength," *Int. J. Adv. Manuf. Technol.*, vol. 89, no. 5–8, pp. 2175–2191, 2017.
16. L. M. Galantucci, F. Lavecchia, and G. Percoco, "Quantitative analysis of a chemical treatment to reduce roughness of parts fabricated using fused deposition modeling," *CIRP Ann. – Manuf. Technol.*, vol. 59, no. 1, pp. 247–250, 2010.
17. L. M. Galantucci, F. Lavecchia, and G. Percoco, "Experimental study aiming to enhance the surface finish of fused deposition modeled parts," *CIRP Ann. – Manuf. Technol.*, vol. 58, n0. 1, pp. 189–192, 2009.
18. A. R. Torrado Perez, D. A. Roberson, and R. B. Wicker, "Fracture surface analysis of 3D-printed tensile specimens of novel ABS-based materials," *J Fail Anal Prev.* vol. 14, no. 3, pp. 343–353, 2014.

19. S. Hwang, E. I. Reyes, K-S. Moon, R. C. Rumpf, and N. S. Kim, "Thermo-mechanical characterization of metal/polymer composite filaments and printing parameter study for fused deposition modeling in the 3D printing process," *J. Electron. Mater.*, vol. 44, no. 3, pp. 771–777 2015.

20. F. Castles *et al.*, "Microwave dielectric characterisation of 3D-printed BaTiO3/ABS polymer composites," *Sci. Rep.*, 6, 22714, 2016.

21. C. M. Shemelya *et al.*, "Mechanical, electromagnetic, and X-ray shielding characterization of a 3D printable tungsten–polycarbonate polymer matrix composite for space-based applications," *J. Electron. Mater.*, vol. 44,no. 8, pp. 2598–2607, 2015.

22. D. V. Isakov, Q. Lei, F. Castles, C. J. Stevens, C. R. M. Grovenor, and P. S. Grant, "3D printed anisotropic dielectric composite with meta-material features," *Mater. Des.*, vol. 93, pp. 423-430, 2016.

23. K. Boparai, R. Singh, and H. Singh, "Comparison of tribological behaviour for Nylon6-Al-Al$_2$O$_3$ and ABS parts fabricated by fused deposition modelling," *Virtual Phys. Prototyp.*, vol. 10, no. 2, pp. 59-66, 2015.

24. M. Nikzad, S. H. Masood, and I. Sbarski, "Thermo-mechanical properties of a highly filled polymeric composites for fused deposition modeling," *Mater. Des.*, vol. 32, no. 6, 3448–3456, 2011.

25. A. R. Torrado, C. M. Shemelya, J. D. English, Y. Lin, R. B. Wicker, and D. A. Roberson, "Characterizing the effect of additives to ABS on the mechanical property anisotropy of specimens fabricated by material extrusion 3D printing," *Addit. Manuf.*, vol. 6, pp. 16–29, 2015.

26. O. S. Carneiro, A. F. Silva, and R. Gomes, "Fused deposition modeling with polypropylene," *Mater. Des.*, Vol. 83, pp. 768–776, 2015.

27. A. N. Dickson, J. N. Barry, K. A. McDonnell, and D. P. Dowling, "Fabrication of continuous carbon, glass and Kevlar fibre reinforced polymer composites using additive manufacturing," *Addit. Manuf.*, vol. 16, pp. 146–152, 2017.

28. M. A. Caminero, J. M. Chacón, I. García-Moreno, and J. M. Reverte, "Interlaminar bonding performance of 3D printed continuous fibre reinforced thermoplastic composites using fused deposition modelling," *Polym. Test.*, vol. 68, pp. 415–423, 2018.

29. F. Van Der Klift, Y. Koga, A. Todoroki, M. Ueda, Y. Hirano, and R. Matsuzaki, "3D Printing of continuous carbon fibre reinforced thermo-plastic (CFRTP) tensile test specimens," *Open J. Compos. Mater.*, vol. 6, no. 01, 18–27, 2016.

30. L. J. Love *et al.*, "The importance of carbon fiber to polymer additive manufacturing," *J. Mater. Res.*, vol. 29, no. 17, pp. 1893–1898, 2014.

31. F. Ning, W. Cong, J. Qiu, J. Wei, and S. Wang, "Additive manufacturing of carbon fiber reinforced thermoplastic composites using fused deposition modeling," *Compos. Part B Eng.*, vol. 80, pp. 369–378, 2015.

32. H. L. Tekinalp *et al.*, "Highly oriented carbon fiber-polymer composites via additive manufacturing," *Compos. Sci. Technol.*, vol. 105, pp. 144–150, 2014.

33. W. Zhong, F. Li, Z. Zhang, L. Song, and Z. Li, "Short fiber reinforced composites for fused deposition modeling," *Mater. Sci. Eng. A*, vol. 301, no. 2, pp. 125–130, 2001.

34. I. Fidan *et al.*, "The trends and challenges of fiber reinforced additive manufacturing," *Int. J. Adv. Manuf. Techno.*, vol. 102, pp. 1801–1818, 2019.

35. J. M. Chacón, M. A. Caminero, E. García-Plaza, and P. J. Núñez, "Additive manufacturing of PLA structures using fused deposition modelling: Effect of process parameters on mechanical properties and their optimal selection," *Mater. Des.*, vol. 124, pp. 143–157, 2017.

36. B. Rankouhi, S. Javadpour, F. Delfanian, and T. Letcher, "Failure analysis and mechanical characterization of 3D printed ABS with respect to layer thickness and orientation," *J. Fail. Anal. Prev.*, vol. 16, pp. 467–481, 2016.

37. W. Wu, P. Geng, G. Li, D. Zhao, H. Zhang, and J. Zhao, "Influence of layer thickness and raster angle on the mechanical properties of 3D-printed PEEK and a comparative mechanical study between PEEK and ABS," *Materials (Basel)*, vol. 8, pp. 5834–5846, 2015.

38. O. A. Mohamed, S. H. Masood, J. L. Bhowmik, M. Nikzad, and J. Azadmanjiri, "Effect of process parameters on dynamic mechanical performance of FDM PC/ABS printed parts through design of experiment," *J. Mater. Eng. Perform.*, vol. 25, no. 7, pp. 2922–2935, 2016.

39. H. G. Lemu and S. Kurtovic, "3D printing for rapid manufacturing: Study of dimensional and geometrical accuracy," in *IFIP Advances in Information and Communication Technology AICT* 384, pp. 470–479, 2012.

40. D. Drummer, S. Cifuentes-Cuéllar, and D. Rietzel, "Suitability of PLA/TCP for fused deposition modeling," *Rapid Prototyp. J.*, vol. 18, no. 6, pp. 500–507, 2012.

41. S. H. Masood, W. Rattanawong, and P. Iovenitti, "Part build orientations based on volumetric error in fused deposition modelling," *Int. J. Adv. Manuf. Technol.*, vol. 16, pp. 162–168, 2000.

42. J. K. Bokam and S. Bhowmik, "Parameters optimization of FDM for the quality of prototypes using an integrated MCDM approach," *IGI Global*, pp. 199–220, 2019.

43. M. Harris, J. Potgieter, R. Archer, and K. M. Arif, "Effect of material and process specific factors on the strength of printed parts in fused filament fabrication: A review of recent developments," *Materials*, vol. 12, no. 10, 1664, 2019.

44. A. W. Gebisa and H. G. Lemu, "Influence of 3D printing FDM process parameters on tensile property of ULTEM 9085," *Procedia Manuf.*, vol. 30, pp. 331–338, 2019.

45. B. Huang, S. Meng, H. He, Y. Jia, Y. Xu, and H. Huang, "Study of processing parameters in fused deposition modeling based on mechanical properties of acrylonitrile-butadiene-styrene filament," *Polym. Eng. Sci.*, vol. 59, no. 1, pp. 120–128, 2019.

46. U. K. uz Zaman, E. Boesch, A. Siadat, M. Rivette, and A. A. Baqai, "Impact of fused deposition modeling (FDM) process parameters on strength of built parts using Taguchi's design of experiments," *Int. J. Adv. Manuf. Technol.*, vol. 101, no. 5–8, pp. 1215–1226, 2019.

47. O. A. Mohamed, S. H. Masood, and J. L. Bhowmik, "Optimization of fused deposition modeling process parameters for dimensional accuracy using I-optimality criterion," *Meas. J. Int. Meas. Confed.*, vol. 81, pp. 174–196, 2016.

48. J. T. Cantrell *et al.*, "Experimental characterization of the mechanical properties of 3D-printed ABS and polycarbonate parts," *Rapid Prototyp. J.*, vol. 3, pp. 89–105, 2017.

49. A. Alafaghani, A. Qattawi, B. Alrawi, and A. Guzman, "Experimental optimization of fused deposition modelling processing parameters: A design-for-manufacturing approach," *Procedia Manuf.*, vol. 10, pp. 791–803, 2017.

50. B. H. Lee, J. Abdullah, and Z. A. Khan, "Optimization of rapid prototyping parameters for production of flexible ABS object," *J. Mater. Process. Technol.*, vol. 169, pp. 54–61, 2005.

51. A. K. Sood, R. K. Ohdar, and S. S. Mahapatra, "Parametric appraisal of mechanical property of fused deposition modelling processed parts," *Mater. Des.*, vol. 89, no. 5–8, pp. 2387–2397, 2010.

52. A. Ouballouch, R. El Alaiji, S. Ettaqi, A. Bouayad, M. Sallaou, and L. Lasri, "Evaluation of dimensional accuracy and mechanical behavior of 3D printed reinforced polyamide parts," in *Procedia Struct. Integr.*, vol. 19, pp. 433–441, 2019.

53. https://www.atome3d.com/collections/tag3d?ls=fr-FR. (Accessed January 08, 2022).

54. http://factory-3d.net/produits/5-Composites. (Accessed January 08, 2022).

55. http://factory-3d.net/produits/1-FilamentsTAGin3D. (Accessed January 08, 2022).

56. "ISO – ISO 12085:1996 – Spécification géométrique des produits (GPS) – État de surface: Méthode du profil – Paramètres liés aux motifs." [Online]. Available: https://www.iso.org/fr/standard/20867.html. (Accessed January 08, 2022).

57. "ISO – ISO 1133-2:2011 – Plastiques – Détermination de l'indice de fluidité à chaud des thermoplastiques, en masse (MFR) et en volume (MVR) – Partie 2: Méthode pour les matériaux sensibles à l'historique temps-température et/ou à l'humidité." [Online]. Available: https://www.iso.org/fr/standard/44274.html. (Accessed January 08, 2022).

58. "ISO – ISO 294-1:2017 – Plastiques – Moulage par injection des éprouvettes de matériaux thermoplastiques – Partie 1: Principes généraux, et moulage des éprouvettes à usages multiples et des barreaux." [Online]. Available: https://www.iso.org/fr/standard/67036.html. (Accessed January 08, 2022).

59. "ISO – ISO 11357-1:2016 – Plastiques – Analyse calorimétrique différentielle (DSC) – Partie 1: Principes généraux." [Online]. Available: https://www.iso.org/fr/standard/70024.html. (Accessed January 08, 2022).

60. "ISO – ISO 1183-1:2012 – Plastiques – Méthodes de détermination de la masse volumique des plastiques non alvéolaires – Partie 1: Méthode par immersion, méthode du pycnomètre en milieu liquide et méthode par titrage." [Online]. Available: https://www.iso.org/fr/standard/55640.html. (Accessed January 08, 2022).

61. http://factory-3d.net/produit/84-VolumicStream30PROMK2. (Accessed January 08, 2022).

62. https://www.imprimante-3d-volumic.com/fr/la-gamme-imprimante-3d-detail/847-stream-30-pro-mk2.cfm. (Accessed January 08, 2022).

63. "Tensile Test Methods for Plastics: ISO 527 and JIS K 7161 : SHIMADZU (Shimadzu Corporation)." [Online]. Available: https://www.shimadzu.com/an/industries/engineering-materials/film/plastics-iso/index.html. (Accessed January 08, 2022).

64. "ISO 4287:1997(fr), Spécification géométrique des produits (GPS) – État de surface: Méthode du profil – Termes, définitions et paramètres d'état de surface." [Online]. Available: https://www.iso.org/obp/ui#iso:std:iso:4287:ed-1:v1:fr. (Accessed January 08, 2022).

65. A. K. Sood, R. K. Ohdar, and S. S. Mahapatra, "Improving dimensional accuracy of fused deposition modelling processed part using grey Taguchi method," *Mater. Des.*, vol. 30, no. 10, pp. 4243–4252, 2009.

66. R. Bansal, "Improving dimensional accuracy of fused deposition modelling (FDM) parts using response surface methodology," p. 32, 2011.

67. M. a Yardimci, S. I. Guceri, M. Agarwala, and S. C. Danforth, "Part quality prediction tools for fused deposition processing," *Solid Free. Fabr. Proceedings, Sept. 1996*, 1996.

68. M. A. M. J. Alhubail, "Statistical-based optimization of process parameters of fused deposition modelling for improved quality," 2012.

69. M. Kaveh, M. Badrossamay, E. Foroozmehr, and A. Hemasian Etefagh, "Optimization of the printing parameters affecting dimensional accuracy and internal cavity for HIPS material used in fused deposition modeling processes," *J. Mater. Process. Technol.*, vol. 226, pp. 280–286, 2015.

70. W. Wu *et al.*, "3D printing of thermoplastic PI and interlayer bonding evaluation," *Mater. Lett.*, vol. 229, pp. 206–209, October 2018.

71. B. V. Reddy, N. V. Reddy, and A. Ghosh, "Fused deposition modelling using direct extrusion," *Virtual Phys. Prototyp.*, vol. 2, no. 1, pp. 51–60, 2007.

72. A. J. Qureshi, S. Mahmood, W. L. E. Wong, and D. Talamona, "Design for scalability and strength optimisation for components created through FDM process," in *Proceedings of the International Conference on Engineering Design, ICED*, vol.6, pp. 255–266, 2015.

73. W. Z. Wu, P. Geng, J. Zhao, Y. Zhang, D. W. Rosen, and H. B. Zhang, "Manufacture and thermal deformation analysis of semicrystalline polymer polyether ether ketone by 3D printing," *Mater. Res. Innov.*, vol. 18, no. S5, pp. S12–S16, 2014.

74. X. Deng, Z. Zeng, B. Peng, S. Yan, and W. Ke, "Mechanical properties optimization of poly-ether-ether-ketone via fused deposition modeling," *Materials (Basel)*, vol. 11, no. 2, 216, 2018.
75. X. Tian, T. Liu, C. Yang, Q. Wang, and D. Li, "Composites : Part A interface and performance of 3D printed continuous carbon fiber reinforced PLA composites," *Compos. Part A*, vol. 88, pp. 198–205, 2016.
76. H. Li, T. Wang, J. Sun, and Z. Yu, "The effect of process parameters in fused deposition modelling on bonding degree and mechanical properties," *Rapid Prototyp. J.*, vol. 24, no. 1, pp. 80–92, 2018.
77. J. Torres, M. Cole, A. Owji, Z. DeMastry, and A. P. Gordon, "An approach for mechanical property optimization of fused deposition modeling with polylactic acid via design of experiments," *Rapid Prototyp. J.*, vol. 22, no. 2, pp. 387–404, 2016.
78. A. Lanzotti, M. Grasso, G. Staiano, and M. Martorelli, "The impact of process parameters on mechanical properties of parts fabricated in PLA with an open-source 3-D printer," *Rapid Prototyp. J.*, vol. 21, no. 5, pp. 604–617, 2015.
79. M. S. Uddin, M. F. R. Sidek, M. A. Faizal, R. Ghomashchi, and A. Pramanik, "Evaluating mechanical properties and failure mechanisms of fused deposition modeling acrylonitrile butadiene styrene parts," *J. Manuf. Sci. Eng.*, vol. 139, no. 8, 2017.
80. B. M. Tymrak, M. Kreiger, and J. M. Pearce, "Mechanical properties of components fabricated with open-source 3-D printers under realistic environmental conditions," *Mater. Des.*, vol. 58, pp. 242–246, 2014.
81. X. Liu, M. Zhang, S. Li, L. Si, J. Peng, and Y. Hu, "Mechanical property parametric appraisal of fused deposition modeling parts based on the gray Taguchi method," *Int. J. Adv. Manuf. Technol.*, vol. 89, no. 5-8, pp. 2387–2397, 2017.
82. F. Knoop and V. Schoeppner, "Mechanical and thermal properties of FDM parts manufactured with polyamide 12," *Solid Free. Fabr. Symp.*, pp. 935–948, 2015.

7 Characterization of Effect of Cellular Support Structures in Selective Laser Melting Using Stainless Steel 316L

M. Abattouy, M. Ouardouz, and H. Azzouzi
Abdelmalek Essaâdi University, Morocco

CONTENTS

7.1 Introduction: Background and Literature Review 143
7.2 Experimental Procedure 146
    7.2.1 Material 146
7.3 Conclusion 155
References 156

7.1 INTRODUCTION: BACKGROUND AND LITERATURE REVIEW

Additive manufacturing (AM) refers to a set of technologies that enable physical components to be created from virtual 3D models by layering the component on top of each other until it is complete [1].

In contrast to subtractive manufacturing, which starts with a block of material and removes any unwanted material (either by carving it by hand or using a machine like a mill, lathe, or CNC machine) until the desired part is left [2, 3], AM starts with nothing and builds the part one layer at a time by "printing" each new layer on top of the previous one until the part is complete. The layer thickness varies depending on the technique utilized, ranging from a few microns to roughly 0.25 mm per layer, and a variety of materials are currently available for the various technologies [4].

The earliest AM concepts date back to the late 19th and early 20th centuries, with the introduction of layer-based topographical maps as 3D representations of terrain, as well as a variety of methods for using these topological models to produce 3D maps, such as wrapping a paper map over the topological models to create a 3D model of the terrain. Photo sculpture, which began at the end of the 19th century and involved taking a series of photographs from various angles around an object, which were then used to carve out the object using each different angled picture as a template [5], making it an initially subtractive process, had several proposed methods

DOI: 10.1201/9780367822385-7

for creating models using photosensitive materials. Modern AM can be traced back to a patent issued by Otto John Munz in 1951, which could be regarded as the beginning of modern stereolithography technology [6]. It was simply a stack of layered 2D transparent images printed on photosensitive emulsions that were piled on top of one another. He devised a method for selectively exposing transparent materials. Each layer of the photo emulsion was exposed with a cross section of an object in a layer-by-layer method. The build platform on which the part was being made was gradually lowered, and the next layer of photo emulsion and fixing agent was created on top of the preceding layer, much like a current stereolithography machine [7]. The output of the printing process was a solid transparent cylinder with a 3D image of the object. The final true three-dimensional (3D) object had to be manually cut or photochemically etched out of the cylinder as a subsequent procedure, which was a flaw in this technology. Swainson proposed a process to directly fabricate a plastic pattern through selective 3D polymerization of a photosensitive polymer at the intersection of two laser beams in 1968 [8], and the following decades saw the development of a succession of new techniques, including those of Swainson, who proposed a process to directly fabricate a plastic pattern through the selective 3D polymerization of a photosensitive polymer at the intersection of two laser beams in 1968 [9].

Photochemical machining [10], in which an object is created by photochemically crosslinking or degrading a polymer through simultaneous exposure to intersecting laser beams, was also done at Battelle Laboratories. Ciraud presented a powder process in 1971 that can be regarded as the forerunner of modern direct deposition AM techniques such as powder bed fusion, while Housholder produced the first powder-based selective laser sintering process in 1979 [11]. He talked about applying planar layers of powder in a sequential manner and selectively hardening portions of each layer in his invention. Heat and either a specified mask or a controlled heat scanning method such as a laser could be used to solidify the material. The development of commercially available systems and commercial AM, as we know it now, did not begin until 1986, with Charles W. Hull's stereolithography patent [12]. UVP Inc. owned the patent at the time, and the company licensed the technology to Charles Hull, a former employee who went on to develop 3D Systems. The first commercial SLA machine was introduced in 1988, and since then, practically every year has seen an exponential increase in the number of systems, technology, and materials available [13].

Even AM vocabulary has evolved significantly during the previous three decades. Because the primary usage of the different available technologies was to build concept models and pre-production prototypes, rapid prototyping (RP) was the primary term used to characterize layer-upon-layer manufacturing technologies for the majority of the 1990s. Solid freeform fabrication (SFF) and layer manufacturing are two more terminologies that have been used over the years.

However, in early 2009, the ASTM International Committee F42 on AM Technologies attempted to standardize the industry's terminology, and after a meeting in which many industry experts debated the best terminology to use, the term "additive manufacturing" was chosen as the industry's standard terminology. In ASTM F2792 10E1 [14] standard terminology for AM technologies document, they described AM as follows:

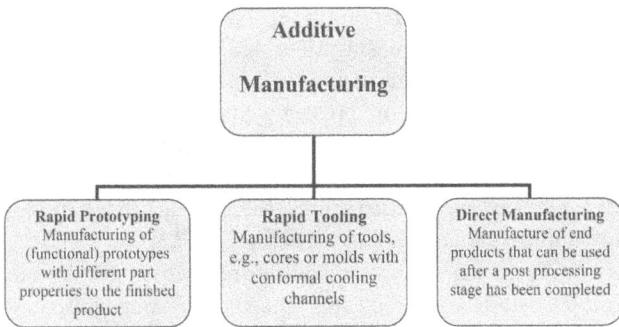

FIGURE 7.1 Terminologies for additive manufacturing [15].

the process of joining materials to make objects from 3D model data, usually layer upon layer, as opposed to subtractive manufacturing methodologies, such as traditional machining.

AM made it possible for small parts, small quantities, and one-off goods, such as those used in the jewelry and medical and dental technology industries, to become increasingly cost-effective to manufacture.

The term "rapid" or "direct" manufacturing was used to describe this method. It brought up entirely new design engineering options, such as the production of geometrically complicated pieces from high-strength materials for lightweight construction. The leading AM companies in the industry agreed in 2010 to replace the multiple application-specific, imprecise, and misleading names prefixed by "rapid" with the new term "additive manufacturing", as illustrated in Figure 7.1.

Unlike subtractive manufacturing, which involves removing material from a larger block of material until the desired result is attained, most AM procedures do not produce a lot of waste. If a part is appropriately "planned for AM", as opposed to a single part manufactured by traditional manufacturing, it may not require the enormous amounts of time required to remove undesired material, potentially saving time and money. This should not be interpreted to mean that AM can always produce cheaper parts than traditional production. Because AM is a rather complex and expensive technology, it is frequently the reverse. However, this is dependent on the AM technology utilized and the numerous design parameters that can be used.

AM is a capable process to produce 3D components from raw materials and 3D design data. This layer-by-layer operating process has many advantages including high geometrical freedom to produce complex parts with reduced cost and applied especially in the aerospace, medical, and automotive industry.

Support structures are necessary for AM [16]; they achieve many technical functions, they increase the ability to manufacture complex geometries, they play a role of heat sink to dissipate the high energy and facilitate heat transfer during part building; as a result, they manage residual stresses which is very interesting for selective laser technology, since high temperature is leading to thermal stresses [17]. Support structures can also keep away part distortion and contribute to providing physical support for weak geometries; in addition, they play a role separating produced part from the build platform for easy removal [18]. The use of support structures affects the cost of manufacture by increasing time and material consumption [19]; they might affect also post-processing cost and surface quality of the part due to removal consequences. Design challenges are present when working with support structures; they must be frangible to achieve an easy removal of parts and at the same time, must be as strong as possible to ensure a successful building [20]. However, even with these advantages of support structures, there exist a small amount of work in literature about optimization of support structure [21–27].

Due to the significant use of aluminum alloy and titanium for the aerospace application, we find that studies are focusing on selective laser melting support structures using titanium and aluminum alloys; however, stainless steel 316L is receiving increased research attention for AM as it exhibits better corrosion resistance and stronger at elevated temperature, and it can be used in heat exchangers, jet engine parts, valve and pump parts, chemical processing equipment, tanks, and evaporators. This work is contributing to achieve the following goals.

A full factorial design of experiments (DOE) is generated for cone support, tree support, and different cellular supports structures manufactured with stainless steel 316L using selective laser melting for selected geometric control factors.

Digital microscopy is used to allow the study of upper surface quality and see through the cross section to study deformation and then a removability evaluation of every sample from the platform is investigated.

7.2 EXPERIMENTAL PROCEDURE

During this experiment, an Nd: YAG laser (Neodymium-doped yttrium aluminum garnet) is used with a wavelength of 1070 ± 10 nm. The laser beam is transported to the scanner optical system through fiber and then transferred to the machine chamber using the SCANLAB scanner. The focal point size is around 70 mm and the laser beam is a continuous laser wave (CW). The research machine's name is Ep-M250 located at the laboratory center for AM at the faculty of sciences and techniques of Tangier, Abdelmalek Essaadi University. The equipment consists of a scanner, laser source, selective laser melting, and chamber that is divided into four main components: powder feed storage container, platform of building part, recycled powder storage container, and a recoating system, Figure 7.2.

The inert gas used in this work, shown in Figure 7.3, is nitrogen supplied from nitrogen generator type Boltec, it can supply nitrogen at the pressure of 0.9 Mpa with a purity of 99.99 percent.

The parameters of the SLM machine used during this study are shown in Table 7.1.

7.2.1 Material

The stainless-steel 316L exhibit better corrosion resistance and stronger at elevated temperature; it can be used in heat exchangers, jet engine parts, valve and pump parts, chemical processing equipment, tanks, and evaporators. Table 7.2 shows the chemical composition. The powder is melted with 30-µm layer thickness during the

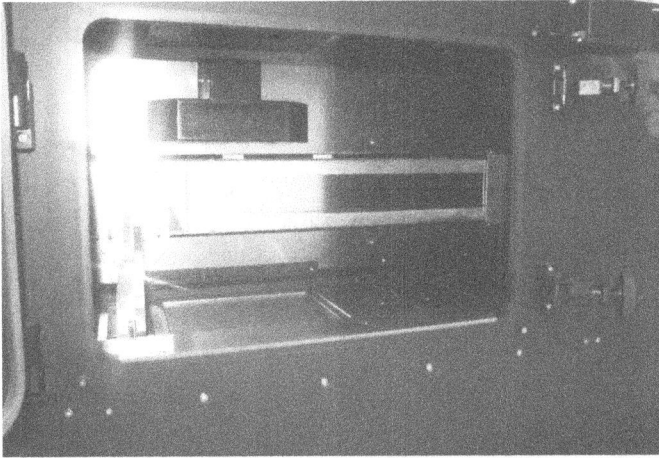

FIGURE 7.2 Building chamber of SLM machine used in this study.

building process. The stainless steel 316L is widely used in aerospace, medical, and other engineering uses that need high resistance of corrosion and strength. With stainless-steel 316L, it can be fabricated spare parts, small series products, functional electromechanical systems, and personalized products.

In this experiment, a prismatic body is designed to create a simplified base for evaluating different support structures, the prismatic body is 18 mm long, 12 mm wide, and 6 mm high. The support structure studied are cone structures, tree structures, and other cellular structures named from A to H, as shown in Figure 7.4, a total of 10 structures have been inspected.

In this set of experiments, cone, tree support, and cell structures were evaluated at two different levels, for the cone and tree support distance between two tubes at

FIGURE 7.3 Inert gas supply of nitrogen.

TABLE 7.1
SLM Machine Parameters

Parameter	Characteristics
Build volume	$262 \times 262 \times 350$ mm³
Layer thickness	0.02–0.1 mm
Laser power	200 or 500 W
Scanning system	High precision galvanometer
Scanning speed	8 m/s

levels 1 and 2 is investigated. For the cellular structures dimension of cell is studied in 3 mm and 6 mm levels as shown in Table 7.3, a total of 20 samples were manufactured and inspected. The figure from CAD data preparation software is included in order to give an understanding of the design matrix.

After manufacturing, the microstructural inspection was carried out and was conducted with digital microscopy in order to investigate the part distortion, surface quality from one hand, from the other hand removability of part was evaluated, the removal of part was carried out with hand saw and evaluated by giving six levels from 0 to 5 and assigned to criteria from very easy to cut to extremely hard to cut. As it can be seen from Table 7.4, a comparison of upper surface quality is established by giving six levels from 0 to 5 (5 for excellent, 4 for good, 3 for average, 2 for bad, 1 for very bad, and 0 for extremely bad).

The top part surface quality of every sample was investigated, Figure 7.7 shows micrographs of 11 samples that have been successfully fabricated, however, the other samples were not given since there was a failure during part building.

TABLE 7.2
Powder Stainless Steel Chemical Composition

Element	Concentration wt.%
C	0.030
Cr	18.00
Cu	0.50
Fe	Balance
Mn	2.00
Mo	5.5
N	0.10
Ni	13
O	0.10
P	0.025
S	0.010
Si	0.75

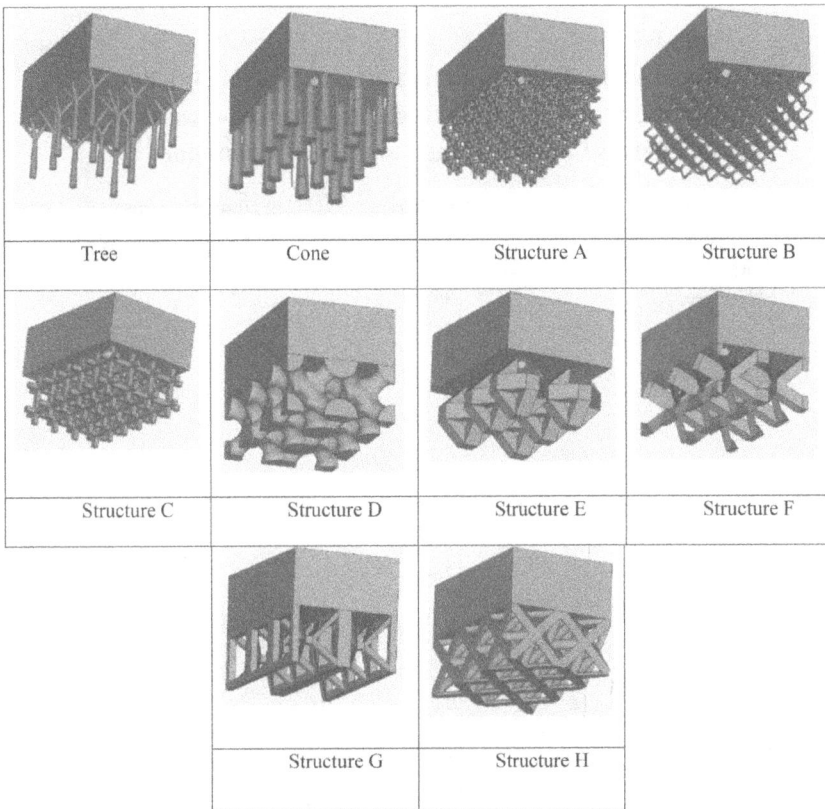

FIGURE 7.4 Three-dimensional view of different support structures tested.

After this overall inspection, test samples were cut from the platform, the cross sections were prepared for the microstructural analysis, and the cross sections were observed compared with digital models in Figure 7.8. It has been shown from the figure that presenting the upper surface and compared with cross section of the part and digital sample shown in Figure 7.5 that sample 8 have the worst overall surface quality and high roughness, this poor overall surface quality might be due to the smaller contact area between the support structure and lower surface of the sample even though there is a higher frequency of support cells of this structure. Samples 5, 10, 13, and 15 show a better overall surface quality; however, there are some deformations of the surface near the part borders, according Figure 7.8 shows the cross sections of these samples, and compared with the digital parts, it's clear that the overhang surface that presents these types of support structure near the borders increased deformations of the upper surface.

From Figures 7.6 and 7.7, it can be noticed that samples 11, 12, 16, 17, 18, and 19 show better upper surface quality in the overall surface as well as in borders, and this might be due to support structure types that have larger contact area between the lower surface of part and support structure as well as it can be seen in the borders

TABLE 7.3

Dimensions of Cell for Every Part and Type of Structure (*for Cone and Tree Structure, the Distance between Two Bars Is Considered Instead of Cell Dimension)

Structure	Part Number	Cell Dimension (mm)
Cone	1	6
	2	3
Tree	3	6
	4	3
C	5	6
	6	3
D	7	6
	8	3
E	9	6
	10	3
F	11	6
	12	3
G	13	6
	14	3
H	15	6
	16	3
I	17	6
	18	3
J	19	6
	20	3

there is no overhang surface due to the support structures geometry that doesn't allow gabs in borders.

The tree support and cone-based support shown to be not able to maintain the building process from the first scanning layers, they are weak structures, and recoating movements affected them and the building is failed. These are shown in Figure 7.6.

The sample N° 5, as s shown in Figure 7.8, the support structure was successfully constructed, but the upper surface of the part is extremely rough, and warping has occurred along the two borders as a result of the gaps not being filled on either side. Because of this overhang surface, the construction process was abandoned during this stage.

Sample N°8 in Figures 7.7 and 7.8 shows a correctly constructed structure, but it was unable to support the weight of its feature because its borders were warped from both sides. The upper surface quality is very low, the heat affected zone was extremely high because of the irregularities, and the downward surface is also rough. As a result, the structure failed to maintain the building.

It appears that the structure N°10 depicted in the Figures N° 7.7 and 7.8 is stronger; nonetheless, warping has occurred on the left side, primarily as a result of wide

TABLE 7.4

Evaluation of Border Distortion, Upper Surface Quality, and Part Removability

Structure	Part Number	Cell Dimension (mm)	Building Situation	Border Deformation	Surface Quality
Cone	1	6	Failed	N/A	N/A
	2	3	Failed	N/A	N/A
Tree	3	6	Failed	N/A	N/A
	4	3	Failed	N/A	N/A
C	5	6	Succeeded	1	0
	6	3	Failed	N/A	N/A
D	7	6	Failed	N/A	N/A
	8	3	Succeeded	0	1
E	9	6	Failed	N/A	N/A
	10	3	Succeeded	2	2
F	11	6	Succeeded	5	5
	12	3	Succeeded	5	5
G	13	6	Succeeded	4	5
	14	3	Failed	N/A	N/A
H	15	6	Succeeded	1	4
	16	3	Succeeded	5	4
I	17	6	Succeeded	5	4
	18	3	Succeeded	5	4
J	19	6	Succeeded	5	5
	20	3	Failed	N/A	N/A

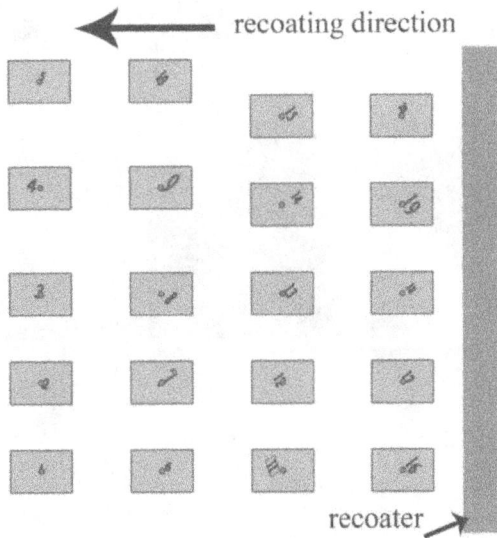

FIGURE 7.5   Digital model of design matrix of the experiment and recoating direction.

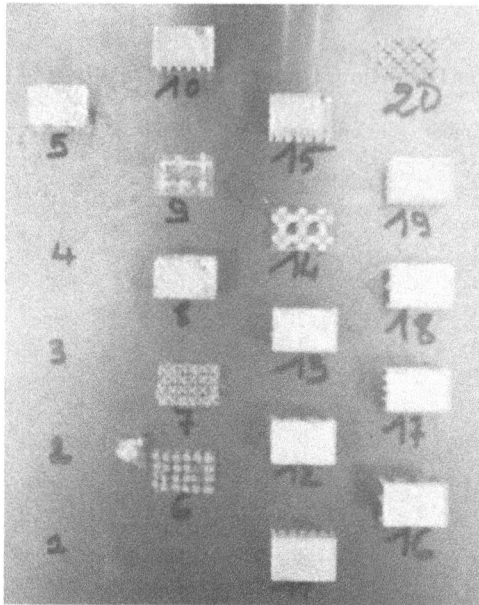

FIGURE 7.6 Upside view of samples fabricated numbered from 1 to 20.

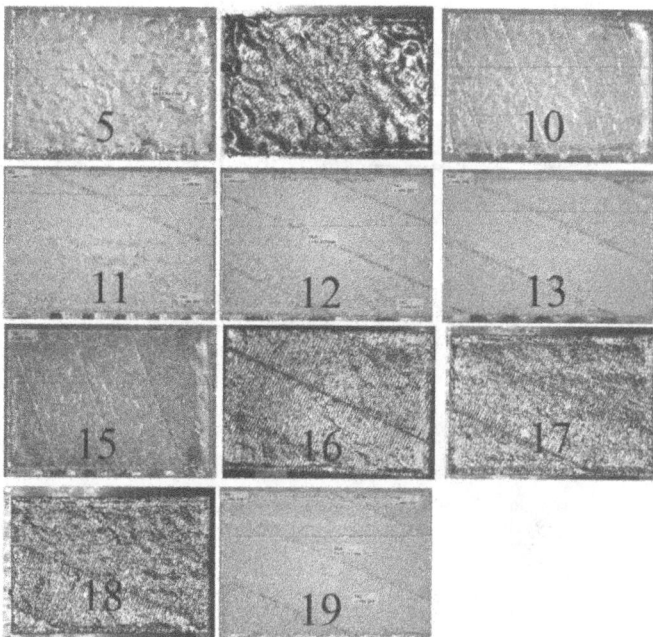

FIGURE 7.7 Upside view of the upper surface of 11 samples fabricate numbered 5, 8, 10–13, and 15–19.

gaps in the cell-based support, which has failed to construct the feature even though the surface quality is superior.

Samples 11, 12, 13, 16, 17, 18, and 19 are included. The building has been successful, and warping has been kept to a minimum. From both sides, these structure supports are distinguished by strong cell features, which fill gaps in the borders and allow for the proper holding of parts in the platform. The quality of both the downward and upward surfaces is significantly higher. These structure supports are acceptable; they may, however, have an impact on part removal from the platform, making it more difficult to remove.

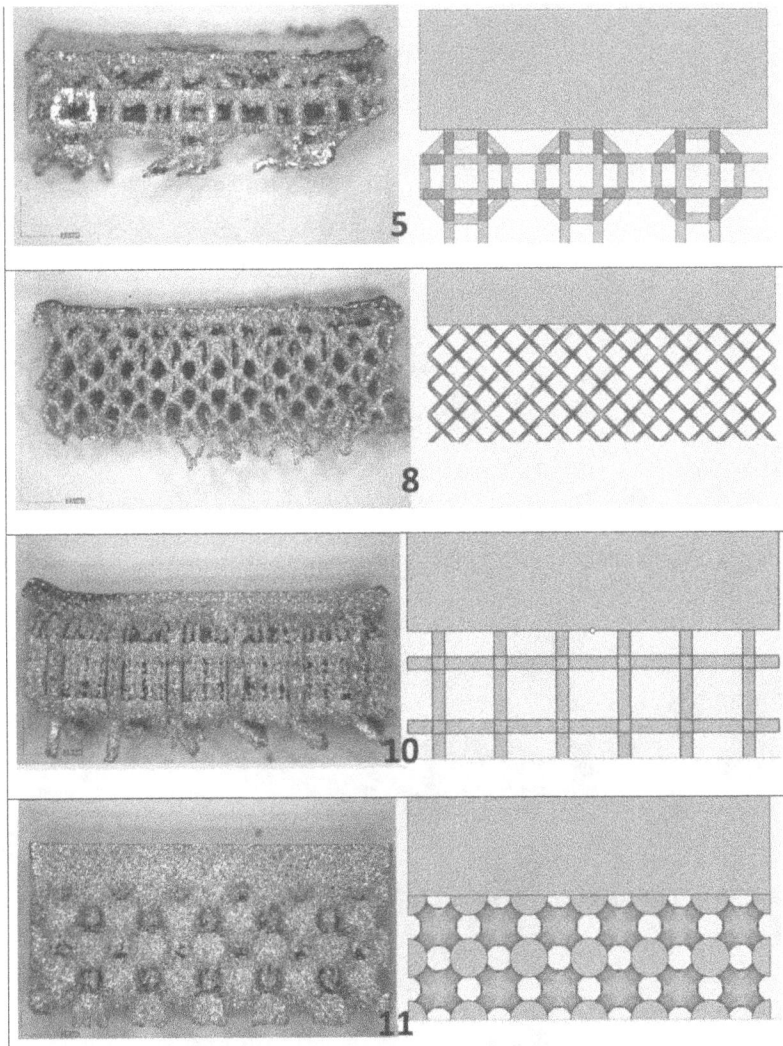

FIGURE 7.8 Cross section view of 11 samples fabricated numbered 5, 8, 10–13, and 15–19 compared with their 3D model *(Continued)*

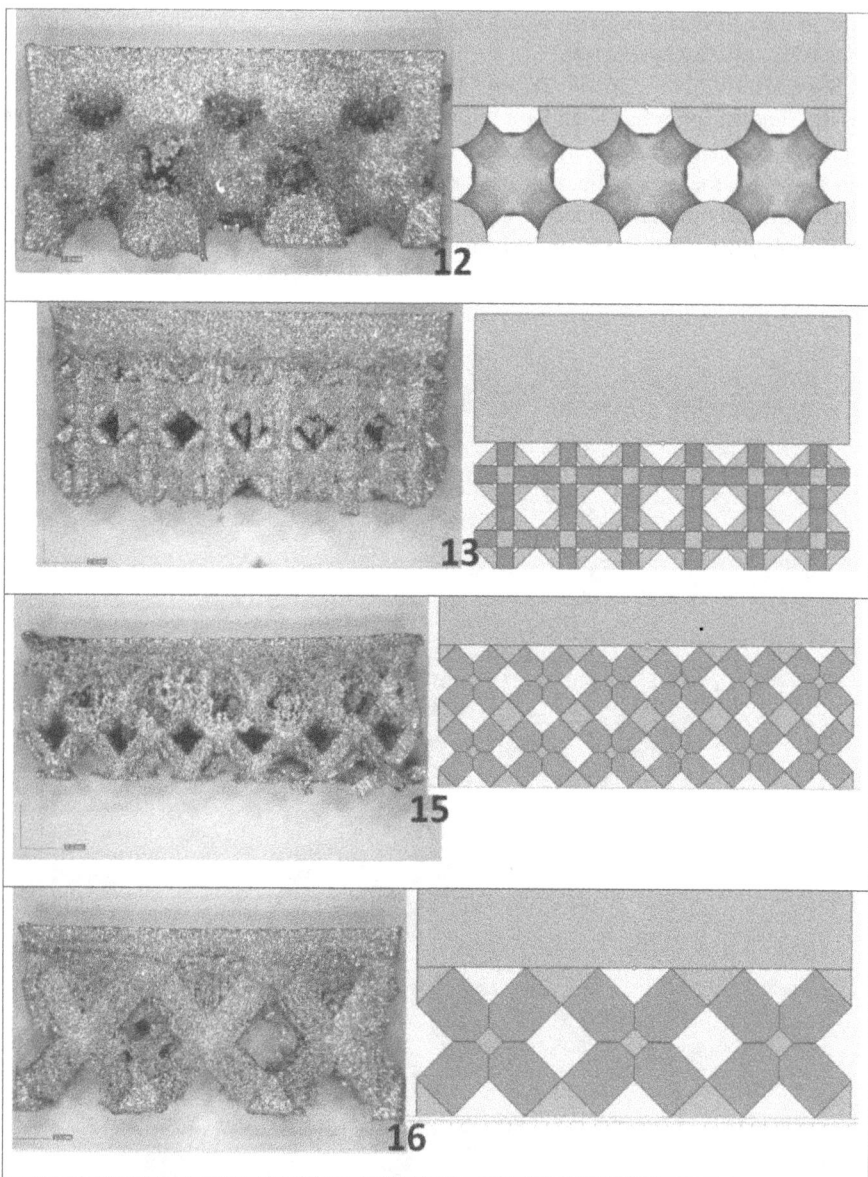

FIGURE 7.8 *(Continued)*

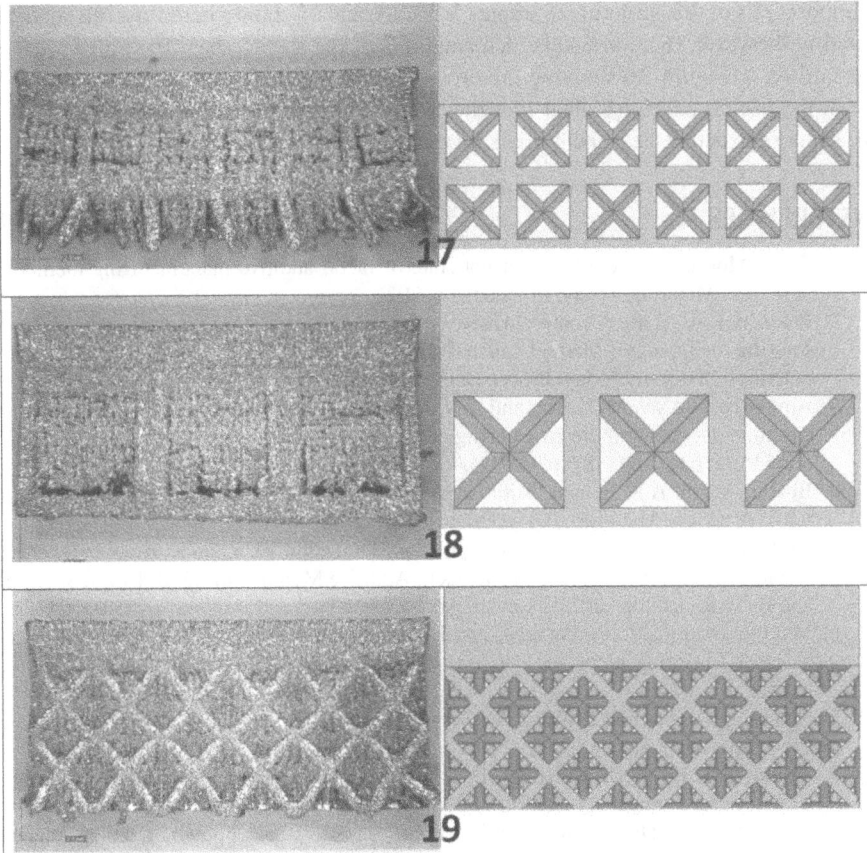

FIGURE 7.8 *(Continued)*

7.3 CONCLUSION

The purpose of this study was to compare the features of two distinct types of support structures (tree and cellular supports). Each support type was manufactured. 316L stainless steel was used as the test material. The experiment employed an IPG YLS 200W SM CW Ytterbium fiber laser with a wavelength of 1070 nm and an EPM 250 SLM laser additive machine. The study's primary focus was on the removability of support structures and the surface quality of the components. The removal job was performed manually, as the removability of the supports should be determined by their ease of removal. The surface properties of every support structure were advantageous during the production process, they must be removed afterward. It has been found that it has been shown from the figure that presenting upper surface and compared with cross section of the part and digital sample shown in Figure 7.5 that sample 8 have the worst overall surface quality and high roughness, this poor overall surface quality might be due to the smaller contact area between the support structure and lower surface of the sample even though there is a higher frequency of support cells of this structure. Samples 5, 10, 13, and 15 show a better overall surface quality; however, there are some deformations of the surface near the part borders, according to Figure 7.6 showing the cross sections of these samples and compared with the digital parts, it's clear that the overhang surface that presents these types of support structure near the borders increased deformations of the upper surface.

REFERENCES

1. Durai Murugan, P., et al., A current state of metal additive manufacturing methods: A review. Materials Today: Proceedings, 2021.
2. Praveen, B.A., et al., A comprehensive review of emerging additive manufacturing (3D printing technology): Methods, materials, applications, challenges, trends and future potential. Materials Today: Proceedings, 2021.
3. Abureden, G.A., W.M. Hasan, and A.N. Ababneh, Exploring potential benefits of additive manufacturing in creating corrugated web steel beams. Journal of Constructional Steel Research, 2021. 187: p. 106975.
4. Bhatia, A. and A.K. Sehgal, Additive manufacturing materials, methods and applications: A review. Materials Today: Proceedings, 2021.
5. Dutta, B., S. Babu, and B. Jared, Chapter 1 – Metal additive manufacturing, in Science, Technology and Applications of Metals in Additive Manufacturing, B. Dutta, S. Babu, and B. Jared, Editors. 2019, Elsevier. pp. 1–10.
6. Mukhtarkhanov, M., A. Perveen, and D. Talamona, Application of stereolithography based 3D printing technology in investment casting. Micromachines, 2020. 11(10): p. 946.
7. Huang, J., Q. Qin, and J. Wang, A review of stereolithography: Processes and systems. Processes, 2020. 8(9): p. 1138.
8. Arefin, A.M.E., et al., Polymer 3D printing review: Materials, process, and design strategies for medical applications. Polymers, 2021. 13(9): p. 1499.
9. Deore, B., et al., Direct printing of functional 3D objects using polymerization-induced phase separation. Nature Communications, 2021. 12(1): p. 55.
10. Allen, D.M., et al., Three-dimensional photochemical machining. CIRP Annals, 1987. 36(1): pp. 91–94.
11. Kumar, S., Selective laser sintering: A qualitative and objective approach. JOM, 2003. 55(10): pp. 43–47.
12. McCausland, T., 3D printing's time to shine. Research-Technology Management, 2020. 63(5): pp. 62–65.
13. Alimi, O.A. and R. Meijboom, Current and future trends of additive manufacturing for chemistry applications: A review. Journal of Materials Science, 2021. 56(30): pp. 16824–16850.
14. F2792-10e1, A., ASTM, Standard Terminology for Additive Manufacturing Technologies. 2012. https://www.nen.nl/en/astm-f2792-10e1-en-1074703
15. Abdulhameed, O., et al., Additive manufacturing: Challenges, trends, and applications. Advances in Mechanical Engineering, 2019. 11(2): p. 1687814018822880.
16. Johannes Lindecke, P.N., et al., Optimization of support structures for the laser additive manufacturing of TiAl6V4 parts. Procedia CIRP, 2018. 74: pp. 53–58.
17. Järvinen, J.-P., et al., Characterization of effect of support structures in laser additive manufacturing of stainless steel. Physics Procedia, 2014. 56: pp. 72–81.
18. Baskett, R., Effects of Support Structure Geometry on SLM Induced Residual Stresses in Overhanging Features. Master thesis in Mechanical Engineering. California Polytechnic State University, San Luis Obispo, CA, USA. 2017, https://doi.org/10.15368/theses.2017.91

19. Al-Ketan, O., D.-W. Lee, and R.K. Abu Al-Rub, Mechanical properties of additively-manufactured sheet-based gyroidal stochastic cellular materials. Additive Manufacturing, 2021. 48: p. 102418.

20. Do, Q.T., C.H.P. Nguyen, and Y. Choi, Homogenization-based optimum design of additively manufactured Voronoi cellular structures. Additive Manufacturing, 2021. 45: p. 102057.

21. Tamijani, A.Y., S.P. Velasco, and L. Alacoque, Topological and morphological design of additively-manufacturable spatially-varying periodic cellular solids. Materials & Design, 2020. 196: p. 109155.

22. Chen, L.-Y., et al., Additive manufacturing of metallic lattice structures: Unconstrained design, accurate fabrication, fascinated performances, and challenges. Materials Science and Engineering: R: Reports, 2021. 146: p. 100648.

23. Gao, J., et al., Topology optimization for multiscale design of porous composites with multi-domain microstructures. Computer Methods in Applied Mechanics and Engineering, 2019. 344: pp. 451–476.

24. Maskery, I., et al., Compressive failure modes and energy absorption in additively manufactured double gyroid lattices. Additive Manufacturing, 2017. 16: pp. 24–29.

25. Nazir, A., et al., A state-of-the-art review on types, design, optimization, and additive manufacturing of cellular structures. The International Journal of Advanced Manufacturing Technology, 2019. 104(9): pp. 3489–3510.

26. Savio, G., et al., Geometric modeling of cellular materials for additive manufacturing in biomedical field: A review. Applied Bionics and Biomechanics, 2018. 2018. https://doi.org/10.1155/2018/1654782

27. Zhang, P., et al., Efficient design-optimization of variable-density hexagonal cellular structure by additive manufacturing: Theory and validation. Journal of Manufacturing Science and Engineering, 2015. 137(2). https://doi.10.1108/RPJ-04-2016-0069

8 Using Carbon Nanotubes for Advanced Manufacturing of Antibiofilm Materials

M. Gomes, R. Teixeira-Santos,
L. C. Gomes, and F. J. Mergulhão
LEPABE – Laboratory for Process Engineering,
Environment, Biotechnology and Energy, Faculty of
Engineering, University of Porto, Porto, Portugal
ALiCE – Associate Laboratory in Chemical Engineering,
Faculty of Engineering, University of Porto, Portugal

CONTENTS

8.1   Introduction 159
8.2   CNT-Based Composites in the Biomedical Field 161
8.3   CNT-Based Composites for Marine Applications 164
8.4   CNT-Based Composites for Water Treatment Applications 166
8.5   CNT-Based Composites for Industrial Applications 168
8.6   CNT-Based Composites and Other Applications 170
8.7   Conclusion 170
Acknowledgments 171
References 171

8.1   INTRODUCTION

Carbon nanotubes (CNTs) are cylindrical tubes of covalently bonded carbon atoms that display extraordinary electronic and mechanical properties. There are two basic types of CNTs: single-wall carbon nanotubes (SWCNTs), which are the fundamental cylindrical structures, and multi-wall carbon nanotubes (MWCNTs), which are made of coaxial cylinders, having interlayer spacing close to that of the interlayer distance in graphite (0.34 nm). These cylindrical structures have a diameter in the order of nanometers (depending on the number of walls) but a length of several microns (100 μm) extendable to up to a few millimeters (about 4 mm) (Upadhyayula and Gadhamshetty 2010). They were first discovered by M. Endo in 1978, but the real interest in CNTs started when the Japanese scientist Iijima first reported them in

1991 (Iijima 1991). The field thrived after that and the first polymer composites using CNTs were reported by Ajayan et al. (1994).

CNTs have very attractive properties, such as high surface area per unit volume, excellent mechanical strength, and high electrical and thermal conductivity (Al-Jumaili et al. 2017; Upadhyayula and Gadhamshetty 2010). The antimicrobial and antifouling (AF) activities of pristine CNTs have also been reported (Chen et al. 2013; Kang et al. 2007). However, the use of CNTs in their bulk form results in a poor translation of their inherent properties, limiting technological advancement (Khan, Sharma, and Saini 2016). To exploit their attributes, it becomes essential to mix CNTs with engineered polymers (e.g., thermoplastics, elastomers, conjugated polymers) or natural polymers to obtain nanocomposites with augmented benefits (Ma et al. 2010; Khan, Sharma, and Saini 2016). Along with the enhancement of the structural, thermal, and electronic properties of the final composites, the conjugation of CNTs with materials such as polymers, metals, or biomolecules frequently results in the production of nanocomposites with increased antimicrobial activity (Teixeira-Santos, Gomes, and Mergulhão 2020).

The incorporation of CNTs in polymer matrices provides materials that can be applied in industrial, environmental, and biomedical fields. Currently, the most widespread use of CNT nanocomposites is in industry. They have been used to produce transistors and chemical sensors (Norizan et al. 2020) and to develop membranes for filtration and other separation processes (Madenli, Yanar, and Choi 2020; Zhang et al. 2017). Additionally, CNTs have been used in the formulation of cleaning agents, biocides, and disinfectants for industrial processes (Vassallo et al. 2018).

In the environmental field, the AF performance and mechanical properties of CNTs allowed their application in marine transportation. The undesired growth of marine organisms on ship hulls decreases their speed and increases fuel consumption, with both economic and environmental penalties. Several studies have proposed the incorporation of CNTs in the synthesis of fouling-release (FR) coatings against microalgae and barnacles (Beigbeder et al. 2008; Martinelli et al. 2011; Cavas et al. 2018). CNTs have also been applied in water and wastewater treatment as absorbents for biological and organic/inorganic contaminants due to significant discoveries related to antimicrobial and adsorption characteristics of CNTs (Smith and Rodrigues 2015; Fan et al. 2019; Cruz-Silva et al. 2019).

More recently, CNTs have been particularly used in pharmacy and medicine as a drug delivery system. It has been demonstrated that the chemical stability of these nanomaterials enables them to adsorb or conjugate with a wide range of therapeutic molecules such as proteins, antibodies, DNA, etc. (He et al. 2013). CNTs have also been used in the construction of biosensors for the detection of biomolecules and biological cells, tissue engineering, and neuronal interfaces (Upadhyayula and Gadhamshetty 2010; He et al. 2013). Due to their antimicrobial activity, CNTs have been studied for the manufacture of medical devices and prosthetic implants (Kim et al. 2019; Cho et al. 2019).

Although CNTs showed to reduce biofilm formation in different scenarios, there is still little consensus about their mode of action. The partial penetration of nanotubes into bacterial cells is pointed out as a potential mechanism to inhibit adhesion. The physical piercing of the outer membranes of microorganisms leads to an irrecoverable membrane perturbation and, consequently, to cell disruption and release of intracellular content (Kang et al. 2008). Moreover, the microtexture of CNTs showed to have a negative impact on bacterial adhesion (Malek et al. 2016). Additionally, it has been suggested that the anti-adhesive effect of MWCNTs is related to their flexibility and oscillation in the substrate, which do not provide a stable surface for the adhesion of microorganisms and consequent biofilm development (Malek et al. 2016).

This chapter focuses on a few recent advanced applications of CNT composites in the domains of industrial processes, environment, and medicine, with particular emphasis on their antimicrobial and AF performance. CNT-based materials will be presented according to their potential final application, and by discussing some studies, the future directions of these nanocomposites in the different fields will be outlined.

8.2   CNT-BASED COMPOSITES IN THE BIOMEDICAL FIELD

CNT-based composite materials have been used in a plethora of biomedical applications, varying from therapeutic (drug and gene delivery, cancer treatment, and tissue engineering and regenerative medicine) to diagnostic applications (biomedical imaging and biosensors) (Raphey et al. 2019; Erol et al. 2018).

Regarding the antimicrobial and anti-adhesive properties of CNT-based materials, they have been exploited not only for designing a variety of drug delivery systems (where their ability to penetrate through the cellular membranes and the high drug carrier capacity are of paramount importance) but also for antimicrobial therapy, and the development of high-performance composites for medical devices and implants.

Table 8.1 describes some representative studies reporting the antibiofilm activity of MWCNT composites with application in the biomedical field. In the last decade, the conjugation of CNTs with compounds displaying antimicrobial activity – antimicrobial agents, photosensitizers, antimicrobial peptides (AMPs), and enzymes – revealed to be a promising approach to prevent biofilm formation (Teixeira-Santos et al. 2021). The favorable combination of MWCNTs with AMPs was particularly highlighted in a study performed by Qi et al. (2011). In this specific case, the covalent immobilization of nisin (a natural AMP with well-known antimicrobial activity) on MWCNTs enhanced the antimicrobial and antibiofilm activities of the final composite, offering new possibilities for the development of novel antimicrobial surfaces.

The functionalization of CNTs with amine and carboxyl groups has also been investigated in this field. It is indeed meant to improve CNT solubility in water and biological fluids and, consequently, reduce CNT cytotoxicity (Anzar et al. 2020). In a study conducted by Zardini et al. (2014), ethanolamine-functionalized MWCNTs proved their efficiency against a broad range of species, presenting higher antimicrobial activity than pristine MWCNTs (p-MWCNTs). According to the authors, this can be explained by the increase in the interaction between amino groups of ethanolamine-functionalized MWCNTs and negatively charged cell walls of microorganisms (Zardini et al. 2014). The influence of MWCNT surface chemistry on bacterial adhesion was also previously assessed by Vagos et al. (2020). However, in this study, p-MWCNTs incorporated in polydimethylsiloxane (PDMS) seemed to be more efficient in the reduction of *Escherichia coli* adhesion compared to the functionalized ones. The authors also reported a significant reduction in cell adhesion by using

TABLE 8.1
Studies Demonstrating the Efficacy of MWCNT Composites with Application in the Medical Field

Application	MWCNT Composite	Species	Major Conclusions	Reference
Antimicrobial surfaces	Nisin immobilized on MWCNTs	E. coli Pseudomonas aeruginosa Staphylococcus aureus Bacillus subtilis	Nisin/MWCNT composites showed higher antimicrobial and antibiofilm activities (7-fold and 100-fold, respectively) than pristine MWCNTs (p-MWCNTs) against all tested pathogens.	Qi et al. (2011)
Antimicrobial agents	Ethanolamine-functionalized MWCNTs (f-MWCNTs)	Gram-positive and -negative bacteria	The antimicrobial activity of f-MWCNTs was higher than p-MWCNTs as demonstrated by minimum inhibitory concentration determination (2.87 ± 0.11–14.22 ± 0.17 µg mL^{-1} vs 6.12 ± 0.16–36.41 ± 0.06 µg mL^{-1}).	Zardini et al. (2014)
Drug delivery	TiO$_2$-Au nanoparticles embedded on MWCNTs	Gram-positive and -negative bacteria Fusarium solani Aspergillus niger Candida albicans	TiO$_2$-Au/MWCNT composites showed high antimicrobial activity against all tested isolates and inhibited Streptococcus pneumoniae, P. aeruginosa, and C. albicans biofilm formation by 75–90%.	Karthika et al. (2018)
Bone and cartilage tissue engineering	Polyethylene glycol (PEG)-functionalized MWCNT/gelatin-chitosan loaded with ciprofloxacin	Gram-positive and -negative bacteria	The antibacterial activity of the drug-loaded gelatin-chitosan/MWCNT-PEG composites was higher than that of drug-loaded gelatin-chitosan composites, as demonstrated by disk diffusion assays (30.80 mm vs 24.93 mm).	Sharmeen et al. (2018)
Dental biomaterials	MWCNTs incorporated in poly(methyl methacrylate) (PMMA)	Streptococcus mutans S. aureus C. albicans	PMMA/MWCNT composites were able to reduce microbial adhesion by 35–95%.	Kim et al. (2019)

(Continued)

TABLE 8.1 (Continued)

Studies Demonstrating the Efficacy of MWCNT Composites with Application in the Medical Field

Application	MWCNT Composite	Species	Major Conclusions	Reference
Medical devices	Silver-plasma polymer fluorocarbon (PPFC) fabricated using MWCNTs	S. aureus Klebsiella pneumoniae	The Ag-PPFC nanocomposites inhibited the bacterial growth of S. aureus and K. pneumoniae by up to 92% and 45%, respectively, compared to uncoated substrates.	Cho et al. (2019)
Urinary tract devices	Pristine (p-BM) and carboxyl (f-BM) ball-milled MWCNT-filled poly(dimethylsiloxane) (PDMS) composites	E. coli	p-BM-MWCNTs and f-BM-MWCNTs showed a reduction in cell adhesion of 42% and 18%, respectively, compared to the same composites without milling; p-BM-MWCNT/PDMS composites reduced E. coli adhesion by 60%.	Vagos et al. (2020)
Wound healing	Heteroatom (N, F, P/B)-doped MWCNTs: NFP/MWCNTs and NFB/MWCNTs	B. subtilis K. pneumoniae E. coli P. aeruginosa	NFP/MWCNT composites inhibited biofilm formation by 73% for B. subtilis, 78% for E. coli, 71% for K. pneumoniae, and 80% for P. aeruginosa; NFB/MWCNTs showed 83%, 81%, 77%, and 77% biofilm inhibition against the above bacteria, respectively.	Murugesan et al. (2020)
Membranes for hemodialysis application	Poly(citric acid)-grafted CNTs (PCA-g-MWCNTs) incorporated as nanofiller in polyethersulfone (PES)	N/A	Compared to commercial PES hemodialysis membranes, the PES/PCA-g-MWCNT membranes showed a lower flux decline (5-fold) and higher water flux recovery ratio (from 15.8 $Lm^{-2}h^{-1}$ to 95.4 $Lm^{-2}h^{-1}$).	Abidin et al. (2016)

N/A – not applicable.

pristine ball-milled MWCNTs (60% reduction), probably driven by a better degree of dispersion in PDMS (Vagos et al. 2020).

The combination of MWCNTs with different formulations of silver, titanium, and copper is also being tested with the intent of producing new drug delivery systems and innovative materials for application on medical devices. In a study performed by Karthika et al. (2018), for instance, MWCNTs decorated with titanium dioxide-gold nanoparticles exhibited great antibiofilm activity against *Shigella dysenteriae, Proteus vulgaris, K. pneumoniae, S. pneumoniae, B. subtilis, S. aureus,* and *C. albicans*. Similarly, Ag-plasma polymer fluorocarbon (PPFC) nanocomposite thin films fabricated using an MWCNT-Ag-polytetrafluoroethylene composite were found to suppress bacterial growth and proliferation by up to 92% (Cho et al. 2019). These results can be attributed not only to the antimicrobial properties of Ag nanoparticles and MWCNTs but also to the super-hydrophobic character of the PPFC matrix (Cho et al. 2019).

As previously mentioned, CNT-based composites have been largely investigated in the tissue engineering and regenerative medicine fields, where they have been used as filler particles in the fabrication of dental and orthopaedical implants (Lekshmi et al. 2020). In addition to their role as strengthening agents, MWCNT-based polymer composites have demonstrated palpable success in the inhibition of bacterial adhesion and biofilm formation either by using poly(methyl methacrylate) (PMMA) (Kim et al. 2019) or PEG/gelatin-chitosan (Sharmeen et al. 2018) polymers.

The association of MWCNT nanocomposites with different heteroatoms (N, F, P/B) also seems to generate promising results, significantly changing the properties of CNTs and extending their potential applications (Murugesan et al. 2020). Heteroatom doping operates in three different ways: it offers the minerals needed for new tissue formation in the wound-healing process; it strongly inhibits the biofilm formation of Gram-negative and -positive strains; and, finally, it increases the biocompatibility and wound healing ability of MWCNT composites (Murugesan et al. 2020).

The incorporation of poly(citric acid)-grafted-MWCNTs as nanofillers in polyethersulfone to produce hemodialysis mixed matrix membranes (MMM) was also addressed by Abidin et al. (2016). Overall, the researchers noticed the enhancement of MMM hydrophilicity, porosity, and AF properties (Abidin et al. 2016).

Although a vast number of studies have already reported the advantages of using MWCNT composites in the medical field, concerns related to their biosafety, toxicity, carcinogenic, and teratogenic effects still exist, limiting their medical application (Lekshmi et al. 2020). Even though functionalization can expand the overall performance of CNTs, promoting their biocompatibility, further investigations are needed to clarify CNT toxicity and thus enable the translation of surface-engineered CNT nanocomposites from the laboratory to the clinic.

8.3 CNT-BASED COMPOSITES FOR MARINE APPLICATIONS

In the last decade, different polymer-based nanocomposites incorporating CNTs as fillers have been investigated for anti-biofouling applications in the marine industry, as seen in Table 8.2. Marine biofouling occurs every time fouling organisms settle and colonize surfaces (of either natural or artificial origin) submerged in a marine

TABLE 8.2

Studies Demonstrating the Efficacy of CNT Composites for Marine Applications

Application	CNT Composite	CNT Type	Type of Study	Species	Major Conclusions	Reference
Marine FR coatings	Silicone-based coatings filled with CNTs and natural sepiolite	MWCNT	In vitro	Representative soft-fouling (Ulva linza) and hard-fouling (Balanus cyprid) organisms	The percentage removal of Ulva spores from the coating containing 0.2% MWCNTs was significantly higher (70%) than that from the other coatings, including the unfilled control. The addition of 0.2% MWCNTs to the PDMS significantly decreased the critical removal stress for barnacles.	Beigbeder et al. (2008)
Coatings for anti-biofouling applications	Carboxyl CNT/PDMS nanocomposites (c-CNT/PDMS)	MWCNT	Sea trials	Early eukaryotic and prokaryotic communities	c-CNT/PDMS surfaces prevented biofouling for more than 14 weeks. c-CNTs significantly reduced eukaryoticmicrobial diversity, which contributes to the hindering of biofilm development.	Sun and Zhang (2016a)
Eco-friendly coatings for marine anti-biofouling applications	Carboxyl and hydroxyl CNT-filled PDMS composites	MWCNT	Sea trials (XiaoShi Island harbor waters)	Early eukaryotic communities	MWCNT- and c-MWCNT/PDMS composites demonstrated exceptional AF properties. The diversity and richness of species of early eukaryotic communities on most CNT-based surfaces were significantly lower than those on PDMS control.	Sun and Zhang (2016b)
		MWCNT	Sea trials (Weihai Western Port, China)	Barnacles, mussels, ascidians, Ulva, seaweeds	The AF properties of PDMS films were greatly improved with the incorporation of a low amount of CNTs (0.1 wt%), causing a strong perturbation on prokaryotic colonization compared to PDMS.	Ji et al. (2018)
		MWCNT SWCNT	Short-term sea trials (XiaoShi Island, near the West Port of the Weihai city, China)	Pioneer surface-biofilm bacteria communities	The incorporation of CNTs in PDMS affected pioneer surface-biofilm bacterial communities. While Proteobacteria decreased on all CNT/PDMS composites compared to PDMS (70 vs 85%), Cyanobacteria increased (22 vs 8%). SWCNT/PDMS composites demonstrated lower AF efficacy compared to MWCNT/PDMS surfaces.	Sun et al. (2020)

environment (Fusetani 2004). As a result of biofouling on ship hulls, some factors are compromised, including its speed and maneuverability and fuel consumption, which tends to increase (Townsin 2003). Given the widely known technical and economic challenges presented by biofouling in the marine industry, the use of CNTs as fillers for polymer composites is becoming increasingly important. By applying these nano-materials, the mechanical strength of the final composite (Cavas et al. 2018) and its ability to retard biofouling – either by AF or FR properties – can be enhanced.

The application of AF and FR coatings is indeed the most common strategy to combat biofouling (Gule, Begum, and Klumperman 2016; Selim et al. 2017). Previously, Beigbeder et al. (2008) have found that the AF and FR properties of the PDMS matrix can be greatly enhanced with the incorporation of a very small amount of MWCNTs (0.05%).

Once again, CNT functionalization plays a crucial role in the development of improved nanocomposites. In fact, recent studies have reported that by incorporating carboxyl- and hydroxyl-modified CNTs in the PDMS matrix exposed to the natural seawater, excellent anti-biofouling properties can be achieved (Sun and Zhang 2016a, b; Ji et al. 2018; Sun et al. 2020). Additionally, it was reported that modified CNT/PDMS composites are able to significantly reduce the diversity and richness of microbial communities, being less prone to the pioneer eukaryotic colonization and the subsequent attachment and colonization of macrofoulers (Sun and Zhang 2016a, b). Although CNTs seem to be good candidates for the preparation of AF and FR coatings, the development of new polymer-modified CNTs for anti-biofouling applications needs to be further addressed. Furthermore, future efforts should focus on the investigation of the anti-biofouling and degradation mechanisms of CNT composites, as well as the production of novel eco-friendly coating materials.

8.4 CNT-BASED COMPOSITES FOR WATER TREATMENT APPLICATIONS

CNT-polymer composites continue to stand out due to their excellent properties, and water and wastewater treatment and filtration are some of the areas in which they have been successfully applied (Sarkar et al. 2018; Kokkinos, Mantzavinos, and Venieri 2020). The excellent performance of CNTs in the adsorption of con-taminants from water derives from their high affinity and selective adsorption capacity for contaminants, in particular, organic pollutants, heavy metals, and bacteria (Kokkinos, Mantzavinos, and Venieri 2020). Some representative studies reporting the efficacy of MWCNT composites for water treatment applications are presented in Table 8.3.

Despite the success of p-CNT meshes in the purification of contaminated drink-ing water (De Volder et al. 2013), there has been a continuous effort towards the functionalization of CNTs. Apart from the modification of CNTs with nisin (Dong and Yang 2015), recent studies have been emphasizing the importance of associating silver and other noble metals with MWCNT/polymer membranes in order to increase their antimicrobial activity and fouling resistance (Gunawan et al. 2011; Macevele, Moganedi, and Magadzu 2017; Pang, Ahmad, and Zaulkiflee 2019; Fan et al. 2019; Ali et al. 2019).

TABLE 8.3

Studies Demonstrating the Efficacy of MWCNT Composites for Water Treatment Applications

Application	MWCNT Composite	Species	Major Conclusions	Reference
Water disinfection and biofouling control	Silver nanoparticles/MWCNTs coated on a polyacrylonitrile (Ag/MWCNT/PAN) hollow fiber membrane	E. coli	The relative flux drop was 6% for the Ag/MWCNT/PAN membranes, being significantly lower than for pristine PAN (55%). The presence of Ag/MWCNTs inhibited bacterial growth and prevented biofilm formation.	Gunawan et al. (2011)
Water disinfection	Nisin-MWCNT coated filters	Bacillus anthracis	Nisin deposited on MWCNT filters increased the B. anthracis capture up to 3.9 log and significantly reduced their viability by 96–97%.	Dong and Yang (2015)
Wastewater treatment	Silver/MWCNT/poly(vinylidene fluoride-co-hexafluoropropene) (Ag/MWCNT/PVDF-HFP) membranes	E. coli	The 3% Ag/MWCNT/PVDF-HFP membranes showed a high fouling resistance rate and bactericidal activity (100% bacterial load reduction).	Macevele, Moganedi, and Magadzu (2017)
Water treatment applications	Polyethyleneimine/MWCNT/trimesoyl chloride (PEI/MWCNT/TMC)	N/A	The hydrophilic and negatively charged PEI/MWCNT/TMC surface generates membranes with good AF properties (90% more than PEI/MWCNT surface).	Liu et al. (2017)
	N-halamine epoxide and siloxane grafted onto the MWCNTs (N/Si/MWCNTs)	E. coli S. aureus	The films containing N/Si/MWCNTs displayed a flux recovery ratio value above 97% and had excellent antibacterial efficacy (98 and 96% against S. aureus and E. coli, respectively).	Huang et al. (2017)
	PES membrane incorporated with zinc oxide (ZnO) and MWCNTs	Enterobacter sp.	ZnO/MWCNT/PES membrane demonstrated efficient AF properties with high flux ratios of 28–56 L m^{-2} h^{-1} versus 7.8 L m^{-2} h^{-1} obtained for PES membrane. Additionally, few bacteria were found attached to the membrane.	Pang, Ahmad, and Zaulkiflee (2019)
	Silver nanoparticles with MWCNTs (Ag/MWCNTs) on ceramic membrane under electrochemical assistance	E. coli	Viable cells on the MWCNT/ceramic membrane were reduced to 2.6 log, while bacteria were completely inactivated by Ag-MWCNT/ceramic membrane.	Fan et al. (2019)
Water and wastewater treatment	MWCNT/polyethylene (MWCNT/PE)	Pseudomonas fluorescens Mycobacterium smegmatis	Biofilm growing on MWCNT/PE surface decreased by 89% and 29% for P. fluorescens and M. smegmatis, respectively, compared to PE surface.	Jing, Sahle-Demessie, and Sorial (2018)
Dairy wastewater treatment	Thermo-responsive N-isopropyle acryleamide (NIPAAm) polymerized on the surface of MWCNTs	N/A	MWCNT-NIPAAm membranes demonstrated a flux recovery ratio of 78–99.9% compared to 47% of PES membranes.	Yaghoubi and Parsa (2019)

N/A – not applicable

The combination of MWCNTs, trimesoyl chloride, and polyethyleneimine has also been reported as a way to improve membrane hydrophilicity and thus increase its AF performance (Liu et al. 2017). Additionally, by using films containing N-halamine epoxide/MWCNTs, an antibacterial effect of 96 and 98% against *E. coli* and *S. aureus*, respectively, was previously achieved (Huang et al. 2017). Similarly, Jing, Sahle-Demessie, and Sorial (2018) showed great antibiofilm activity of MWCNT/polyethylene surfaces against *Pseudomonas fluorescens* (biofilm growth was reduced by 89%). Some of these films also present a high water flux recovery rate compared to pristine membranes (Huang et al. 2017; Yaghoubi and Parsa 2019). In this regard, MWCNT/polymer composites are considered to have a promising future for use in next-generation filtration membranes.

In spite of all these successful results, some novel approaches to composite membrane formation using CNTs need to be assessed to achieve maximum contaminant removal efficiency, which includes, for instance, the introduction of other types of functionalization. Still, some other problems need to be evaluated, such as environmental and human exposure to CNTs, scale-up, and CNTs leaching. Further investigations are required to evaluate the ability of functionalized CNT composite membranes to treat water and wastewater in real conditions, which implies long-term applications, as well as to standardize the production procedure of these membranes, facilitating the commercialization for full-scale water treatment.

8.5   CNT-BASED COMPOSITES FOR INDUSTRIAL APPLICATIONS

Over the past decade, CNTs have attracted enormous attention in the industrial field. They have been used to produce chemical sensors and emission transistors, as well as membranes for filtration and separation/purification processes (Zhang et al. 2017; Madenli, Yanar, and Choi 2020). As a consequence, there has been a growing interest in the development of CNT composites with highly antimicrobial and AF activity. Table 8.4 highlights some of the studies that have been carried out in this area.

Tiraferri, Vecitis, and Elimelech (2011) reported that SWCNTs covalently binding to polyamide membranes could efficiently prevent bacterial attachment and, consequently, delay membrane biofouling, being useful for filtration processes. In addition, the immobilization of silver nanoparticles (AgNPs)/MWCNTs with polymer colloids (using a new strategy consisting of a sandwiched type structure) revealed great antimicrobial activity against *E. coli* and *S. aureus* (Rusen et al. 2014).

Later, it has been shown that it is possible to enhance the AF properties of composite membranes by taking advantage of their inherent conductivity and using different forms of applied electrochemical stimulation. In this regard, interlaced CNT electrodes (ICE) on a commercial polyvinylidene fluoride microfiltration membrane showed AF properties at direct (DC) and alternating current (AC) electric potentials through filtration and backwash cycles (Zhang et al. 2017). Also, the authors have stated that (i) ICE AF mechanism can be related with the dielectrophoretic effect in the solution between electrodes, and it is better accomplished at AC in comparison with DC; and (ii) electrochemistry changes bacterial morphology, however, it presents less effect on bacterial density than electrokinetics (Zhang et al. 2017). Recently, Madenli, Yanar, and Choi (2020) revealed the promising antibacterial and

TABLE 8.4

Studies Demonstrating the Efficacy of CNT Composites for Industrial Processing Applications

Application	CNT Composite	CNT Type	Species	Major Conclusions	Reference
Membrane-based separation applications	CNTs covalently bound to polyamide membranes	SWCNT	E. coli	SWCNT membranes achieved up to 60% inactivation of the attached bacteria after 1 h of contact. Additionally, SWCNTs delayed the onset of membrane biofouling during operation.	Tiraferri, Vecitis, and Elimelech (2011)
Filtration processes	Sandwiched-type structure based on polymer colloids, CNTs, and AgNPs	MWCNT	E. coli S. aureus	The polymer colloids/AgNPs/MWCNT exhibited good antimicrobial activity as demonstrated by the disk inhibition zone (11.5 and 9.7 mm for E. coli and S. aureus, respectively, vs ≈ 7.0 mm obtained for the control).	Rusen et al. (2014)
Microfiltration in industrial processing	Interlaced CNT electrodes on a polyvinylidene fluoride microfiltration membrane	MWCNT	P. fluorescens	The optimal operating conditions (2 V alternating current) reduced the fouling rate by 75% compared to control and achieved up to 96% fouling resistance recovery.	Zhang et al. (2017)
Separation and purification applications	CNTs blended PES membranes	MWCNT	E. coli P. aeruginosa	No E. coli colonies were observed on the composite membranes with 0.5% MWCNTs. When the membranes were incubated with P. aeruginosa suspensions, they showed almost 87% less biofilm formation at 24 h compared to PES membrane.	Madenli, Yanar, and Choi (2020)

antibiofilm properties of MWCNT/polyethersulfone membranes and indicated their potential to be used in separation and purification processes.

All the above studies highlight the potential of CNT membranes to enhance bio-fouling resistance compared to the conventional polymeric substrates, recommending their use for industrial processing applications.

8.6 CNT-BASED COMPOSITES AND OTHER APPLICATIONS

Given their outstanding properties, in recent years, CNTs have been used in other applications, including the construction of chemical sensors and the development of cleaning agents, biocides, and disinfectants.

Graphene oxide/MWCNT/poly(O-toluidine) composite was synthesized to fabricate a sensitive and selective chemical sensor for the detection of Pb^{2+} ions in the environment. Antimicrobial studies demonstrated that this composite was effective against both *B. subtilis* and *E. coli* bacteria compared to the amoxicillin as demonstrated by disk diffusion (21 and 8 mm, respectively) and MIC determination (45 and 60 $\mu g\ mL^{-1}$, respectively) methods (Khan et al. 2016).

The antimicrobial properties of some nanomaterials have created interest in their use as cleaning agents, biocides, and disinfectants. Vassallo et al. (2018) evaluated the antimicrobial activity of MWCNTs and obtained high MIC values (>100 mg L^{-1}), suggesting that this nanomaterial displays low toxicity against *E. coli* cells, although several authors have demonstrated the efficacy of MWCNTs against a broad range of bacteria (Zhang et al. 2015; Hartono et al. 2018). It is known that the percentage of inactivated cells is influenced by the CNT concentration and can be increased by their modification or association with polymers, metals, or biomolecules (Hartono et al. 2018; Upadhyayula and Gadhamshetty 2010). Therefore, the antimicrobial activity of CNTs depends on a multiplicity of factors that may be modulated according to the desired application.

8.7 CONCLUSION

CNTs are described as excellent nanomaterials for numerous applications, including inhibiting biofilm formation. Although the CNT mechanism of action is still under discussion, their antimicrobial and AF activities seem to depend on a variety of factors, which may be tuned in order to improve their efficacy. The functionalization of CNTs is also essential to increase their hydrophilicity and, consequently, the biocompatibility required for medical and environmental applications. According to the analyzed studies, there are several materials such as polymers, biomolecules, and metals, that may be blended to develop effective CNT-based nanocomposites.

The high antimicrobial activity of CNT-nanocomposites was reported against a broad spectrum of microorganisms and their potential for medical and water treatment applications was demonstrated. Also, the significant fouling resistance of these nanocomposites was proven at distinct levels, including in the development of marine AF or FR coatings, water treatment, and industrial processes such as filtration. Nevertheless, further studies are needed to validate the efficacy of CNT-based composites and to translate these findings from the laboratory to real scenarios.

ACKNOWLEDGMENTS

This work was financially supported by: LA/P/0045/2020 (ALiCE), UIDB/00511/2020 and UIDP/00511/2020 (LEPABE), funded by national funds through FCT/MCTES (PIDDAC), and by Project PTDC/CTMCOM/4844/2020 funded by FCT (the Portuguese Foundation for Science and Technology). M. Gomes acknowledges the receipt of a PhD grant from FCT (2021.07149.BD). R. Teixeira-Santos acknowledges the receipt of a junior researcher fellowship from the Project PTDC/BII-BIO/29589/2017 - POCI-01-0145-FEDER-029589. L. Gomes acknowledges FCT for the financial support of her work contract through the Scientific Employment Stimulus – Individual Call – [CEECIND/01700/2017].

REFERENCES

Abidin, M. N. Z., P. S. Goh, A. F. Ismail, M. H. D. Othman, H. Hasbullah, N. Said, Shsa Kadir, F. Kamal, M. S. Abdullah, and B. C. Ng. 2016. Antifouling polyethersulfone hemodialysis membranes incorporated with poly (citric acid) polymerized multi-walled carbon nanotubes. *Materials Science and Engineering C: Materials for Biological Applications* 68:540–550.

Ajayan, P. M., O. Stephan, C. Colliex, and D. Trauth. 1994. Aligned carbon nanotube arrays formed by cutting a polymer resin – nanotube composite. *Science* 265:1212–1214.

Al-Jumaili, A., S. Alancherry, K. Bazaka, and M. V. Jacob. 2017. Review on the antimicrobial properties of carbon nanostructures. *Materials* 10:1066–1091.

Ali, S., S. A. U. Rehman, H. Y. Luan, M. U. Farid, and H. Huang. 2019. Challenges and opportunities in functional carbon nanotubes for membrane-based water treatment and desalination. *Science of the Total Environment* 646:1126–1139.

Anzar, N., R. Hasan, M. Tyagi, N. Yadav, and J. Narang. 2020. Carbon nanotube – a review on synthesis, properties and plethora of applications in the field of biomedical science. *Sensors International* 1:100003–100012.

Beigbeder, A., P. Degee, S. L. Conlan, R. J. Mutton, A. S. Clare, M. E. Pettitt, M. E. Callow, J. A. Callow, and P. Dubois. 2008. Preparation and characterisation of silicone-based coatings filled with carbon nanotubes and natural sepiolite and their application as marine fouling-release coatings. *Biofouling* 24:291–302.

Cavas, L., P. Gokfiliz Yildiz, P. Mimigianni, A. Sapalidis, and S. Nitodas. 2018. Reinforcement effects of multiwall carbon nanotubes and graphene oxide on PDMS marine coatings. *Journal of Coatings Technology and Research* 15:105–120.

Chen, H., B. Wang, D. Gao, M. Guan, L. Zheng, H. Ouyang, Z. Chai, Y. Zhao, and W. Feng. 2013. Broad-spectrum antibacterial activity of carbon nanotubes to human gut bacteria. *Small* 9:2735–2746.

Cho, E., S. H. Kim, M. Kim, J.-S. Park, and S.-J. Lee. 2019. Super-hydrophobic and anti-microbial properties of Ag-PPFC nanocomposite thin films fabricated using a ternary carbon nanotube-Ag-PTFE composite sputtering target. *Surface and Coatings Technology* 370:18–23.

Cruz-Silva, R., Y. Takizawa, A. Nakaruk, M. Katouda, A. Yamanaka, J. Ortiz-Medina, A. Morelos-Gomez, S. Tejima, M. Obata, K. Takeuchi, T. Noguchi, T. Hayashi, M. Terrones, and M. Endo. 2019. New insights in the natural organic matter fouling mechanism of polyamide and nanocomposite multiwalled carbon nanotubes-polyamide membranes. *Environmental Science & Technology* 53:6255–6263.

De Volder, M., S. Tawfick, R. Baughman, and A. J. Hart. 2013. Carbon nanotubes: present and future commercial applications. *Science* 339:535–539.

Dong, X., and L. Yang. 2015. Dual functional nisin-multi-walled carbon nanotubes coated filters for bacterial capture and inactivation. *Journal of Biological Engineering* 9:20–29.

Erol, O., I. Uyan, M. Hatip, C. Yilmaz, A. B. Tekinay, and M. O. Guler. 2018. Recent advances in bioactive 1D and 2D carbon nanomaterials for biomedical applications. *Nanomedicine* 14:2433–2454.

Fan, X., Y. Liu, X. Wang, X. Quan, and S. Chen. 2019. Improvement of antifouling and antimicrobial abilities on silver–carbon nanotube based membranes under electrochemical assistance. *Environmental Science & Technology* 53:5292–5300.

Fusetani, N. 2004. Biofouling and antifouling. *Natural Product Reports* 21:94–104.

Gule, N. P., N. M. Begum, and B. Klumperman. 2016. Advances in biofouling mitigation: a review. *Critical Reviews in Environmental Science and Technology* 46:535–555.

Gunawan, P., C. Guan, X. Song, Q. Zhang, S. S. Leong, C. Tang, Y. Chen, M. B. Chan-Park, M. W. Chang, K. Wang, and R. Xu. 2011. Hollow fiber membrane decorated with Ag/MWNTs: toward effective water disinfection and biofouling control. *ACS Nano* 5:10033–10040.

Hartono, M. R., A. Kushmaro, X. Chen, and R. S. Marks. 2018. Probing the toxicity mechanism of multiwalled carbon nanotubes on bacteria. *Environmental Science and Pollution Research* 25:5003–5012.

He, H., L. A. Pham-Huy, P. Dramou, D. Xiao, P. Zuo, and C. Pham-Huy. 2013. Carbon nanotubes: applications in pharmacy and medicine. *BioMed Research International* 2013:578290–578301.

Huang, Y.-W., Z.-M. Wang, X. Yan, J. Chen, Y.-J. Guo, and W.-Z. Lang. 2017. Versatile polyvinylidene fluoride hybrid ultrafiltration membranes with superior antifouling, antibacterial and self-cleaning properties for water treatment. *Journal of Colloid and Interface Science* 505:38–48.

Iijima, S. 1991. Helical microtubules of graphitic carbon. *Nature* 354:56–58.

Ji, Y., Y. Sun, Y. Lang, L. Wang, B. Liu, and Z. Zhang. 2018. Effect of CNT/PDMS nanocomposites on the dynamics of pioneer bacterial communities in the natural biofilms of seawater. *Materials* 11:902–913.

Jing, H., E. Sahle-Demessie, and G. A. Sorial. 2018. Inhibition of biofilm growth on polymer-MWCNTs composites and metal surfaces. *Science of The Total Environment* 633:167–178.

Kang, S., M. Herzberg, D. F. Rodrigues, and M. Elimelech. 2008. Antibacterial effects of carbon nanotubes: size does matter! *Langmuir* 24:6409–6413.

Kang, S., M. Pinault, L. D. Pfefferle, and M. Elimelech. 2007. Single-walled carbon nanotubes exhibit strong antimicrobial activity. *Langmuir* 23:8670–8673.

Karthika, V., P. Kaleeswarran, K. Gopinath, A. Arumugam, M. Govindarajan, N. S. Alharbi, J. M. Khaled, M. N. Al-anbr, and G. Benelli. 2018. Biocompatible properties of nano-drug carriers using TiO$_2$-Au embedded on multiwall carbon nanotubes for targeted drug delivery. *Materials Science and Engineering: C* 90:589–601.

Khan, A. A. P., A. Khan, M. M. Rahman, A. M. Asiri, and M. Oves. 2016. Lead sensors development and antimicrobial activities based on graphene oxide/carbon nanotube/poly(O-toluidine) nanocomposite. *International Journal of Biological Macromolecules* 89:198–205.

Khan, W., R. Sharma, and P. Saini. 2016. Carbon nanotube-based polymer composites: synthesis, properties and applications. In *Carbon Nanotubes-Current Progress of Their Polymer Composites*, edited by Mohamed Reda Berber and Inas Hazzaa Hafez. Croatia: InTech.

Kim, K.-I., D.-A. Kim, K. D. Patel, U. S. Shin, H.-W. Kim, J.-H. Lee, and H.-H. Lee. 2019. Carbon nanotube incorporation in PMMA to prevent microbial adhesion. *Scientific Reports* 9:4921-4932.

Kokkinos, P., D. Mantzavinos, and D. Venieri. 2020. Current trends in the application of nanomaterials for the removal of emerging micropollutants and pathogens from water. *Molecules* 25:2016–2046.

Lekshmi, G., S. S. Sana, V. H. Nguyen, T. H. C. Nguyen, C. C. Nguyen, Q. V. Le, and W. Peng. 2020. Recent progress in carbon nanotube polymer composites in tissue engineering and regeneration. *International Journal of Molecular Science* 21:6440–6454.

Liu, Y., Y. Su, J. Cao, J. Guan, L. Xu, R. Zhang, M. He, Q. Zhang, L. Fan, and Z. Jiang. 2017. Synergy of the mechanical, antifouling and permeation properties of a carbon nanotube nanohybrid membrane for efficient oil/water separation. *Nanoscale* 9:7508–7518.

Ma, P.-C., N. A. Siddiqui, G. Marom, and J.-K. Kim. 2010. Dispersion and functionalization of carbon nanotubes for polymer-based nanocomposites: a review. *Composites Part A: Applied Science and Manufacturing* 41:1345–1367.

Macevele, L. E., K. L. M. Moganedi, and T. Magadzu. 2017. Investigation of antibacterial and fouling resistance of silver and multi-walled carbon nanotubes doped poly(vinylidene fluoride-co-hexafluoropropylene) composite membrane. *Membranes (Basel)* 7:35–48.

Madenli, E. C., N. Yanar, and H. Choi. 2020. Enhanced antibacterial properties and suppressed biofilm growth on multi-walled carbon nanotube (MWCNT) blended polyethersulfone (PES) membranes. *Journal of Environmental Chemical Engineering* 9:104755–104763.

Malek, I., C. F. Schaber, T. Heinlein, J. J. Schneider, S. N. Gorb, and R. A. Schmitz. 2016. Vertically aligned multi walled carbon nanotubes prevent biofilm formation of medically relevant bacteria. *Journal of Materials Chemistry B* 4:5228–5235.

Martinelli, E., M. Suffredini, G. Galli, A. Glisenti, M. E. Pettitt, M. E. Callow, J. A. Callow, D. Williams, and G. Lyall. 2011. Amphiphilic block copolymer/poly(dimethylsiloxane) (PDMS) blends and nanocomposites for improved fouling-release. *Biofouling* 27:529–541.

Murugesan, B., N. Pandiyan, K. Kasinathan, A. Rajaiah, M. Arumuga, P. Subramanian, J. Sonamuthu, S. Samayanan, V. R. Arumugam, K. Marimuthu, C. Yurong, and S. Mahalingam. 2020. Fabrication of heteroatom doped NFP-MWCNT and NFB-MWCNT nanocomposite from imidazolium ionic liquid functionalized MWCNT for antibiofilm and wound healing in Wistar rats: synthesis, characterization, in-vitro and in-vivo studies. *Materials Science and Engineering: C* 111:110791–110810.

Norizan, M. N., M. H. Moklis, S. Z. Ngah Demon, N. A. Halim, A. Samsuri, I. S. Mohamad, V. F. Knight, and N. Abdullah. 2020. Carbon nanotubes: functionalisation and their application in chemical sensors. *RSC Advances* 10:43704–43732.

Pang, W. Y., A. L. Ahmad, and N. D. Zaulkiflee. 2019. Antifouling and antibacterial evaluation of ZnO/MWCNT dual nanofiller polyethersulfone mixed matrix membrane. *Journal of Environmental Management* 249:109358–109367.

Qi, X., G. Poernomo, K. Wang, Y. Chen, M. B. Chan-Park, R. Xu, and M. W. Chang. 2011. Covalent immobilization of nisin on multi-walled carbon nanotubes: superior antimicrobial and anti-biofilm properties. *Nanoscale* 3:1874–1880.

Raphey, V. R., T. K. Henna, K. P. Nivitha, P. Mufeedha, C. Sabu, and K. Pramod. 2019. Advanced biomedical applications of carbon nanotube. *Materials Science and Engineering C: Materials for Biological Applications* 100:616–630.

Rusen, E., A. Mocanu, L. C. Nistor, A. Dinescu, I. Călinescu, G. Mustăţea, Ş. I. Voicu, C. Andronescu, and A. Diacon. 2014. Design of antimicrobial membrane based on polymer colloids/multiwall carbon nanotubes hybrid material with silver nanoparticles. *ACS Applied Materials & Interfaces* 6:17384–17393.

Sarkar, B., S. Mandal, Y. F. Tsang, P. Kumar, K. H. Kim, and Y. S. Ok. 2018. Designer carbon nanotubes for contaminant removal in water and wastewater: a critical review. *Science of the Total Environment* 612:561–581.

Selim, M. S., M. A. Shenashen, Sherif A. El-Safty, S. A. Higazy, M. M. Selim, H. Isago, and A. Elmarakbi. 2017. Recent progress in marine foul-release polymeric nanocomposite coatings. *Progress in Materials Science* 87:1–32.

Sharmeen, S., A. F. M. Mustafizur Rahman, M. M. Lubna, K. S. Salem, R. Islam, and M. A. Khan. 2018. Polyethylene glycol functionalized carbon nanotubes/gelatin-chitosan nanocomposite: an approach for significant drug release. *Bioactive Materials* 3:236–244.

Smith, S. C., and D. F. Rodrigues. 2015. Carbon-based nanomaterials for removal of chemical and biological contaminants from water: a review of mechanisms and applications. *Carbon* 91:122–143.

Sun, Y., Y. Lang, Z. Y. Yan, L. Wang, and Z. Zhang. 2020. High-throughput sequencing analysis of marine pioneer surface-biofilm bacteria communities on different PDMS-based coatings. *Colloids and Surfaces B: Biointerfaces* 185:110538–110545.

Sun, Y., and Z. Zhang. 2016a. Anti-biofouling property studies on carboxyl-modified multi-walled carbon nanotubes filled PDMS nanocomposites. *World Journal of Microbiology and Biotechnology* 32:148–158.

Sun, Y., and Z. Zhang. 2016b. New anti-biofouling carbon nanotubes-filled polydimethylsiloxane composites against colonization by pioneer eukaryotic microbes. *International Biodeterioration & Biodegradation* 110:147–154.

Teixeira-Santos, R., M. Gomes, L. C. Gomes, and F. J. Mergulhão. 2021. Antimicrobial and anti-adhesive properties of carbon nanotube-based surfaces for medical applications: a systematic review. *iScience* 24:102001–102022.

Teixeira-Santos, R., M. Gomes, and F. J. Mergulhão. 2020. Carbon nanotube-based antimicrobial and antifouling surfaces. In *Engineered Antimicrobial Surfaces*, edited by S. Snigdha, Sabu Thomas, E. K. Radhakrishnan and Nandakumar Kalarikkal, 65–93. Singapore: Springer Singapore.

Tiraferri, A., C. D. Vecitis, and M. Elimelech. 2011. Covalent binding of single-walled carbon nanotubes to polyamide membranes for antimicrobial surface properties. *ACS Applied Materials & Interfaces* 3:2869–2877.

Townsin, R. L. 2003. The ship hull fouling penalty. *Biofouling* 19:9–15.

Upadhyayula, V. K. K., and V. Gadhamshetty. 2010. Appreciating the role of carbon nanotube composites in preventing biofouling and promoting biofilms on material surfaces in environmental engineering: a review. *Biotechnology Advances* 28:802–816.

Vagos, M. R., M. Gomes, J. M. R. Moreira, O. S. G. P. Soares, M. F. R. Pereira, and F. J. Mergulhão. 2020. Carbon nanotube/poly(dimethylsiloxane) composite materials to reduce bacterial adhesion. *Antibiotics* 9:434–448.

Vassallo, J., A. Besinis, R. Boden, and R. D. Handy. 2018. The minimum inhibitory concentration (MIC) assay with *Escherichia coli*: an early tier in the environmental hazard assessment of nanomaterials? *Ecotoxicology and Environmental Safety* 162:633–646.

Yaghoubi, Z., and J. B. Parsa. 2019. Preparation of thermo-responsive PNIPAAm-MWCNT membranes and evaluation of its antifouling properties in dairy wastewater. *Materials Science and Engineering: C* 103:109779–109786.

Zardini, H. Z., M. Davarpanah, M. Shanbedi, A. Amiri, M. Maghrebi, and L. Ebrahimi. 2014. Microbial toxicity of ethanolamines–multiwalled carbon nanotubes. *Journal of Biomedical Materials Research Part A* 102:1774–1781.

Zhang, Q., P. Arribas, E. M.Remillard, M. C. García-Payo, M. Khayet, and C. D. Vecitis. 2017. Interlaced CNT electrodes for bacterial fouling reduction of microfiltration membranes. *Environmental Science & Technology* 51:9176–9183.

Zhang, Q., J. Nghiem, G. J. Silverberg, and C. D. Vecitis. 2015. Semiquantitative performance and mechanism evaluation of carbon nanomaterials as cathode coatings for microbial fouling reduction. *Applied and Environmental Microbiology* 81:4744–4755.

9 Development of a Novel Nanocomposite Coating for Tribological Applications

Arti Yadav, Muthukumar M., and M. S. Bobji
Indian Institute of Science, Bangalore, India

CONTENTS

9.1 Introduction 175
9.2 Manufacturing of Nanocomposite Coating 177
9.3 Formation of Nanoporous Alumina (NPA)-Based
     Nanocomposite Coatings 178
    9.3.1 Anodisation.. 178
    9.3.2 Formation of Nanocomposite by Electrodeposition 180
        9.3.2.1 DC Deposition 181
        9.3.2.2 AC Electrodeposition.. 181
        9.3.2.3 Pulse Electrodeposition 182
9.4 Tribological Characterisation of NPA-Based Nanocomposite Coatings 184
    9.4.1 Wear Mechanisms 186
        9.4.1.1 Wear Track on Porous Alumina... 186
9.5 Conclusion 190
References.. 191

9.1 INTRODUCTION

Nanocomposite coating is composed of mixing two or more dissimilar materials at the nanoscale to control and form a new and improved structure. The properties of nanocomposites coatings and/or materials depend not only on the materials used but also on the morphology and the interfacial characteristics [1–3]. Nanocomposite coatings and/or materials are one of the most exciting and fastest-growing areas of research with new materials and novel properties being continuously developed, which are previously unknown in the constituent materials. Because of their unique properties and increasing popularity, nanostructured and composite coatings have been the subject of numerous books, journal articles, reports, research papers, etc. Therefore, nanocomposite materials and coatings have enormous potential for new applications emerging in various areas including aerospace, automotive, electronics,

DOI: 10.1201/9780367822385-9

biomedical implants, lightweight materials, nanowires, sensors, batteries, bio-ceramics, energy conversion and many other manufacturing applications. This chapter is intended to provide an overview of recent developments of highly uniform and inexpensive nanocomposite coatings. This will also be on the fundamental tribological mechanisms that control their superior friction and wear properties at severe operation conditions (high surface loads and temperatures, lack of lubricants, and aggressive oxidising environments).

In general, hard coatings are widely used for many decades in tribological applications to protect various tools and parts from wear. The development of wear-resistant coatings started in the 1960s. Chemical vapour deposition (CVD) and physical vapour deposition (PVD) techniques are two major techniques that are widely used in many industrial fields that not only provide dramatic improvements in terms of productivity but also have high hardness and corrosion resistance [4–6]. However, the main hurdle is to select the best coating is not often straightforward, as the quality of proper hard coating materials, their combinations, and their deposition process. Other synthesising techniques such as anodising, electroplating, etc., for protecting the surface or forming a protective nanostructured layer/or composite are constantly increasing. During the time of 1980s and since then, materials such as nitrides, carbides, oxides, carbon-based, borides [7–11] are widely being used in hard coating technology due to their outstanding mechanical and tribological properties that provide high hardness and wear resistance, excellent chemical stability and oxidation resistance in severe environments.

For engineering applications, high hardness must be complemented with high fracture toughness. High fracture toughness is necessary when a concentrated load is applied for applications such as high contact load applications. When the coating substrate materials deform significantly than a tough material with high strength provides good elastic recovery. However, the introduction of ductility in hard materials is very challenging. Poros anodised alumina (NPA) is one of the most widely used wear-resistant ceramics to protect the surface [12]. The coating structure can be amorphous or exhibit several crystalline phases, which generally show higher hardness [12]. The structure obtained strongly depends upon the anodising process parameters, and thus, the film properties can vary considerably. Since NPA has excellent hardness and wear resistance, further modification of NPA pores filled with metal can be complemented with high hardness as a next-generation coating.

It is obvious that future tribosystems will be subjected to much more stringent operation conditions than before, mainly because of the increased power density dissipated (e.g. cutting tools) or transmitted (e.g. gears and bearings) at mechanical interfaces and because of the trend towards reduced size and much higher mechanical and thermal loadings at the contact area [13, 14]. To overcome these challenges, new coatings are urgently needed with a capacity to further improve durability and performance and to adapt to the much harsher and rapidly changing operating conditions of future mechanical systems. Accordingly, in the following sections, we provide a brief overview of recent developments in the design and deposition of nanostructured and composite coatings. We also review the tribological properties of such coatings that are important for their performance in various applications.

9.2 MANUFACTURING OF NANOCOMPOSITE COATING

Conventional composite materials consist of one or more materials with different phases and depending on the phases; the discontinuous phase is called the matrix, which is usually harder and has superior mechanical properties, whereas the continuous phase is called reinforcement or reinforcing materials.

Based on the nature of matrix, nanocomposites can be divided into three main groups as follows

- Ceramic-matrix nanocomposites
- Metal-matrix nanocomposites
- Polymer-matrix nanocomposites

Nanocomposite coating depending on their matrix nanocomposites can be prepared by different techniques and the deposition method is usually chosen based on the required coating application and desired coating properties. There are several methods such as sol-gel method, cold spray method, CVD method, PVD method, thermal spray method, electrodeposition, in-situ polymerisation method, spray coating and spin coating methods and dip coating are used. For high deposition rate and uniform deposition in the case of complicated geometries, the CVD, PVD and electrodeposition methods are advantageous and widely used matured techniques compared to other methods.

PVD is an excellent technology for nanoparticle encapsulation in ceramic matrix. In industries and at a commercial scale, the availability of functionalised and non-agglomerating nanoparticles would further allow for significant improvements in nanocomposite manufacturing. However, it must be pointed out that PVD coatings represent a reliable and cost-effective technique for the improvement of tools and machine parts, but it is much complicated and increase in cost when a relatively small dimension coating or nanocomposite coatings are manufactured [15].

CVD is the process that uses a thermally induced chemical reaction between a volatile compound of a material to be added with other gases and produces a non-volatile material that deposits on the appropriate substrate [16]. However, in the case of CVD processes, because the temperatures are typically much higher, CVD may not be a good choice for many heat-sensitive substrates [17]. If a low-temperature CVD is used, a postdeposition hardening heat treatment is often a must, but depending on the coating type, major problems may occur on the hard coating itself. Chief among them is severe oxidation or partial delamination of coatings from the substrate surface due to chemical reactions and thermal distortions.

In electrodeposition, the deposition of a pure metal or its alloy from an electrolyte chemical solution occurs on the cathode when an external current or potential is applied. Electrodeposited coatings are commonly used to protect surfaces from wear and corrosion, as well as for decorative purposes [18, 19]. In this technique, an electrically conductive substrate is required. Nanostructured metal coatings of pure metals, alloys and composites have widely been used in industries using the electrodeposition technique. Using this technique, we can also produce porous coatings filled subsequent densification such as nanoporous alumina (NPA) coating.

The nanopores can be filled with metal like nickel, copper, silver etc., and provide a highly protective nanocomposite coating [12, 20]. Therefore, we focus on NPA-based nanocomposite coating using the electrodeposition technique where NPA is formed by adonisation.

9.3 FORMATION OF NANOPOROUS ALUMINA (NPA)-BASED NANOCOMPOSITE COATINGS

9.3.1 ANODISATION

Anodisation is a surface treatment process commonly used to form a protective oxide coating on the surface of metals like aluminium. Anodised coatings show excellent adhesion characteristics but are porous and brittle. The porosity of the coating reduces the hardness, and the brittle nature of the oxide induces cracking. In practice, the pores are typically filled with organic dye and sealed. Under certain controlled electrochemical conditions, anodisation results in a highly ordered hexagonal porous structure in pure aluminium.

Porous alumina formation can be carried out in either galvanostatic or potentiostatic mode depending on whether applied current density or the applied potential is kept constant, respectively. The variation of voltage and current in galvanostatic and potentiostatic mode while anodising in 20% H_2SO_4 at 1°C is shown in Figure 9.1 [3]. In the galvanostatic mode, initially, voltage increases with time (stage a), representing gradual growth of the dense oxide layer. As the thickness of this compact layer increases, nucleation of pores happens on its surface (stage b). The potential reached a maximum and as the pores propagate into the oxide layer (stage c) starts decreasing. Eventually, a steady-state voltage is reached, and the thickness of the generated ordered porous alumina keeps growing. In potentiostatic mode (Figure 9.1b), the initial high current density decreases to a minimum as the pores are formed in the compact oxide layer and reach a steady-state value during the steady growth of the ordered film.

The two basic chemical processes that take place during the anodisation of aluminium are oxidation of aluminium and the dissolution of the formed alumina [22]. The aluminium ions form and get uniformly distributed at the aluminium–oxide interface.

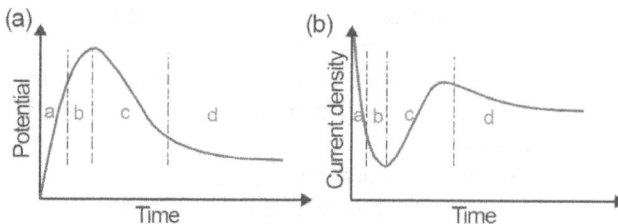

$$Al \rightarrow Al^{3+} + 3e^- \qquad (9.1)$$

FIGURE 9.1 Schematic of the voltage and current transient in (a) galvanostatic and (b) potentiostatic modes [3, 21].

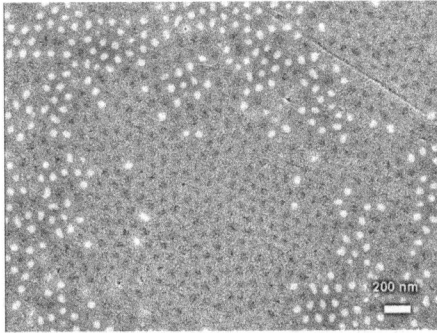

FIGURE 9.2 DC deposited surface showing pores filled with copper revealed after polishing.

These aluminium ions get oxidised either by oxygen ions or hydroxyl ions.

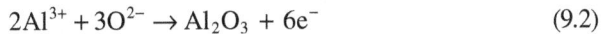

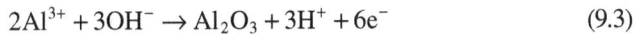

$$2Al^{3+} + 3O^{2-} \rightarrow Al_2O_3 + 6e^- \qquad (9.2)$$

$$2Al^{3+} + 3OH^- \rightarrow Al_2O_3 + 3H^+ + 6e^- \qquad (9.3)$$

Either the aluminium cation has to diffuse through interstitial in the oxide layer to the electrolyte [23] or the anions (oxygen or hydroxide) have to diffuse through the layer to the aluminium metal (Figure 9.2). The necessary oxygen and or hydroxyl ions are supplied from the electrolyte by a water-splitting reaction.

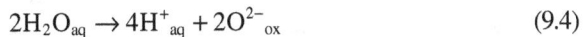

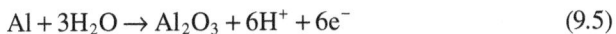

$$2H_2O_{aq} \rightarrow 4H^+{}_{aq} + 2O^{2-}{}_{ox} \qquad (9.4)$$

$$Al + 3H_2O \rightarrow Al_2O_3 + 6H^+ + 6e^- \qquad (9.5)$$

Negative O^{2-} ions can also form from the absorption of OH^- ions at the electrolyte interface [23–25]. Under steady-state, the rate of these reactions determines the thickness and morphology of the resulting oxide film.

In near-neutral solutions of boric acid, ammonium borate, ammonium tartrate or ammonium tetraborate, the rate of water-splitting reaction (Eq. 9.4) is slow. As the compact oxide layer grows, the aluminium, oxygen and hydroxyl ions have to transit through the existing oxide layer. The currently held view is that all the three ions move through the oxide [26, 27]. The movement of these ions is governed by a high field conduction equation,

$$j = A \exp (B\ E) \qquad (9.6)$$

where j is the current density contributed by any one of the ion's transport across the oxide film of thickness t under an external applied field E. The electric field can be approximated to

$$E = U / t \qquad (9.8)$$

where U is the potential across the film. In galvanostatic mode, the voltage will keep on increasing as the oxide layer thickness (t) increases, such that a constant electric field of the order 10^6 to 10^7 V/cm [26, 28]. This will ultimately lead to dielectric breakdown. In potentiostatic mode, since the voltage applied is constant, the electric field across the oxide layer will gradually drop as the oxide thickness increases. The corresponding current density and hence the transport of the ions through oxide drops (Figure 9.1b). The oxidation will eventually stop when the oxide film thickness reaches a critical value (t_c) [27, 29].

Once the pores are formed in the oxide layer, the local electric field is enhanced due to the reduced thickness of the insulating oxide layer. An enhanced electric field at the bottom of the pores results in an oxide-metal interface assuming the shape of the pore [27]. Getting highly ordered NPA on an Al substrate then can be obtained using a two-step anodisation process that was first introduced by Masuda et al. [30]. The geometrical parameters of porous alumina such as pore diameter, interpore distance and film thickness can be altered by controlling the anodising parameters such as anodising voltage, electrolyte concentration, temperature and duration of anodisation. After the anodisation process, two layers, a porous layer and a non-conducting dense oxide layer called the barrier layer (BL) form. The BL exists at the interface between the porous alumina and aluminium [31]. Since the BL is non-conducting, it prevents the formation of porous alumina-based nanocomposites coating via electrochemical deposition of metal [29, 32–36]. To fill the pores with metal, the BL needs to be thinned or removed. Several studies have shown that the BL can be removed or thinned either by cathodic polarisation [37–39] or stepwise reduction in anodisation voltage/current [40–45]. However, cathodic polarisation not only etches the BL but also etches the pore walls thickness of PAA [39]. After thinning of the BL by such a process, metal can be filled into the pore by electrodeposition process and formed nanocomposites.

9.3.2 FORMATION OF NANOCOMPOSITE BY ELECTRODEPOSITION

For depositing metal into the pores electrochemically, there has to be a conducting path through the oxide layer. Metal filling by electrodeposition into porous alumina is usually performed by two processes. One is removing the aluminium completely by etching and deposition of conducting layer on the reverse side, but it is difficult to handle ultrathin oxide layer [45]. The second process involves removing the BL from the bottom of the pores completely, but it is difficult to obtain uniformity over the entire surface [37, 45]. We follow the second process to get the electrochemical conditions at the bottom of the pores as uniform as possible. However, even after the BL thinning, the main challenge was to achieve uniform filling of all the pores simultaneously.

Depending on the type of voltage signal used, the electrodeposition of metals into the pores can be carried out by three different methods.

- DC deposition
- AC deposition
- Pulse deposition

9.3.2.1   DC Deposition

Figure 9.2 shows the top view SEM micrograph of porous alumina after filling copper by the DC electrodeposition method. The surface of the alumina was seen scattered with tiny pyramids of copper. The size of the pyramids varied from a few nanometres to a few micrometres. To determine the amount of pore filling of copper into the samples, they were mechanically polished gently on a selvyt cloth. It can be seen from the SEM image (Figure 9.2) that most of the pores are empty. In the DC deposition method, it was found that we had very little control over the process.

9.3.2.2   AC Electrodeposition

Without separation or removing BL electrodeposition of metal in porous alumina is possible under ac electrodeposition. When an AC voltage or current is applied to the BL, it acts as a rectifier allowing current preferentially in one direction [32]. This is possible because of the fact that electrons can move through the BL with much ease compared to bigger ions like Al^{3+} or O^- and OH^- ions. Deposition of metals like nickel, cobalt, cadmium, bismuth, iron, silver and gold in porous alumina using ac electrodeposition has been reported [46]. However, AC electrodeposition through the BL is a complicated process because the deposition depends on the frequency of the signal applied [47, 48]. To optimise the metal deposition and effect of deposition frequency for ac electrodeposition of copper in porous alumina, a systematic study has been performed.

The uniform growth of metal by ac electrodeposition at the voltage-controlled mode with a continuous 250 Hz, 500 Hz and 750 Hz frequency sine wave was used under 17 V. Figure 9.3a shows the sinusoidal voltage applied and the corresponding current (Figure 9.3b) flowing through the electrochemical cell. It can be seen that the current is more or less symmetric about the x-axis indicating that the total effective charge transferred per cycle is very small. The effect of ac electrodeposition on porous alumina during copper filling is characterised using SEM.

Figure 9.4 shows the SEM image of the top surface of porous alumina after filling copper into porous alumina. When a continuous sine wave of applied voltage 17 V ac at 250 Hz frequency was applied (Figure 9.4a), It was seen that the porous alumina

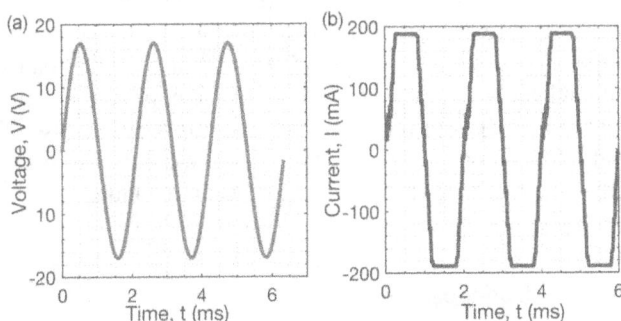

FIGURE 9.3 (a) Applied voltage transient sine wave during AC electrodeposition. (b) Current trace of sine waves during AC electrodeposition.

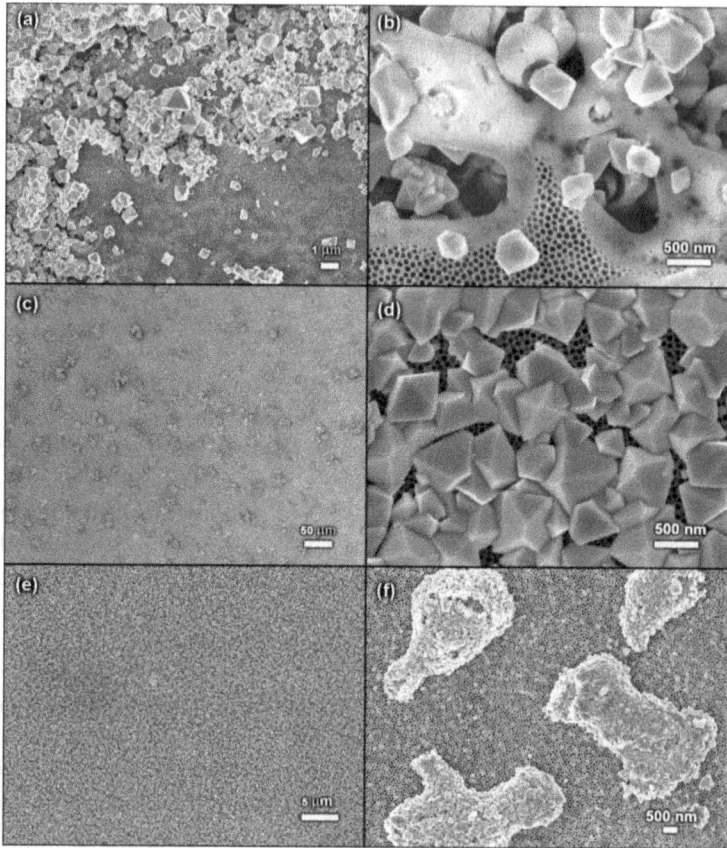

FIGURE 9.4 SEM images of porous alumina filled with copper using AC electrodeposition at various frequencies of (a) 250 Hz, (b) 500 Hz and (c) 750 Hz. Images (d), (e) and (f) are zoom-in details of corresponding SEM images (a), (b) and (c), respectively.

surface was damaged in a few places. The operating frequency of applied voltage was increased to 500 Hz, numerous pyramids were observed on the top of the surface. The size of pyramids varied from ~200 nm to ~800 nm over the surface (Figure 9.4b). When the frequency was maintained at 700 Hz, it was seen that bunches of pyramids were accumulated together and formed an island of pyramids. The surface was found damaged at the edges of these pyramids. It might be because of aluminium oxide/hydroxide dissolution producing hydrogen gas, as discussed by Zhao et al. [39]. Therefore, the evolution of hydrogen gas is becoming more dominant, resulting in damage to the porous alumina layer.

9.3.2.3 Pulse Electrodeposition

It is identified that damage occurs on the porous alumina layer under continuous ac electrodeposition conditions. To avoid this effect, instead of a continuous ac signal, the pulse electrodeposition based on a step square wave with relaxing voltage is used for filling. The square wave signal gives a better pore filling compared to the sine waveform [29, 47, 49]. The alternating voltage pulse signal in the millisecond range with a delay is used for copper filling into porous alumina.

The pulse electrodeposition with the delay time seems to be an effective method for filling metal [29, 36]. The polarisation voltage of 7 V, making the sample positive, was applied for 3.2 ms, followed by a negative voltage of 17 V applied for 3.2 ms with the applied voltage pulse with the delay time. Switch in voltage values occurred over a few microseconds. Deposition of copper happened during this negative pulse. A delay time of 50 ms was given before the next cycle. This delay helps in the restoration of copper ion concentration at the pore bottom before the next pulse and prevents excessive hydrogen evolution [49, 50]. This improves the homogeneity of deposition. The delay should not be more than 2 s because the dissolution reaction between the oxide and electrolyte can become dominant [29, 49].

Figure 9.5 shows the current traces measured at the beginning of the deposition. As the positive pulse is applied, the current increase to 40 mA and as the voltage is held constant at 7 V for 3.2 ms, the current decreases to a value of about 10 mA. The positive pulse is applied to discharge the capacitance of the BL. When this positive cycle switches to the negative cycle, a sudden jump in the current (250 to −230 mA) occurs. During each pulse of negative current, copper is deposited at the pore bottom. At the beginning of the delay time, a sharp increment of current around 70 mA occurs for a few microseconds and afterwards current decreases immediately and reaches zero. The time delay allows for the diffusion of ions to take place from the bulk to the pore. The copper filled into porous alumina samples was examined by SEM, as shown in Figure 9.6. From the cross-sectional images (Figures 9.6c and 9.6e), it is shown that the copper is filled into the pores. The top surface is the image after removing the overgrowth shown in Figure 9.6d, it can be seen that almost the entire surface is filled with copper nanostructures. Cross-section and the surface image after mechanical polishing to remove overgrowth further confirm this.

The ordered porous alumina layer is formed by a two-step anodisation process. By optimising the anodisation conditions, the thickness of the coating and the pore

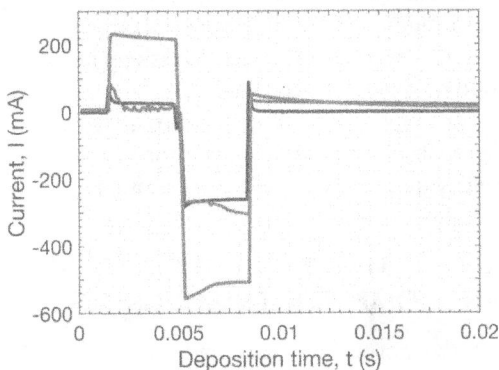

FIGURE 9.5 Current trace of three different square pulses during pulse electrodeposition.

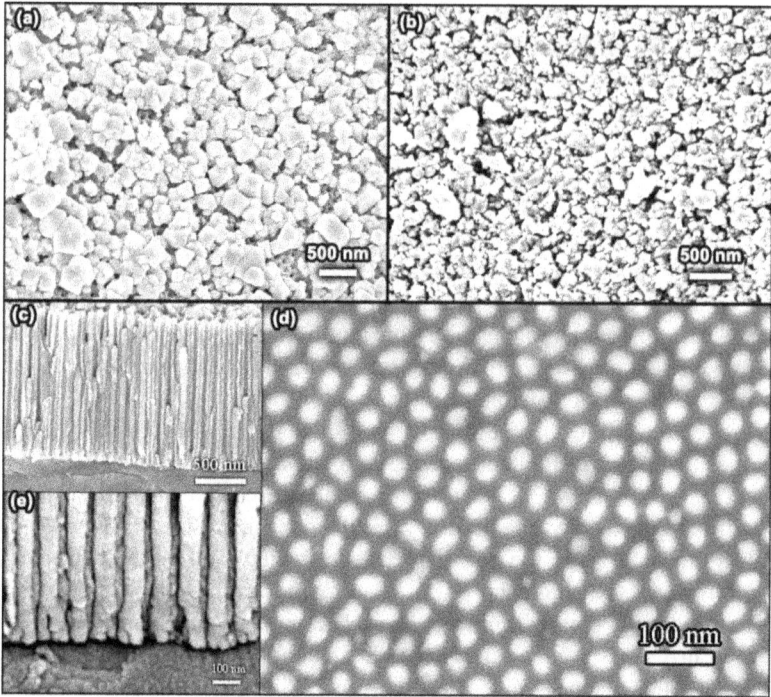

FIGURE 9.6 SEM images of porous alumina using pulse electrodeposition (a) and (b) are the top views, (c) and (e) are the cross-section images of porous alumina after copper deposition and (d) top view of deposited copper on porous alumina after polishing.

size. The interface of the porous structure and aluminium substrate is defined by a non-conducting dense barrier oxide layer. However, to deposit metal into the pores, a conducting path should be established through the BL. One possibility is to etch out the bottom of the pores. But this will reduce the interface strength of the coating. To prevent this, we tried to create a dendritic structure in the BL by gradual reduction of voltage towards the end of anodisation. Optimisation of the dendritic structure led to the uniform deposition of metal into pores, achieved by pulsed electrodeposition. In pulse electrodeposition, a positive pulse is applied to remove accumulated charge near the bottom of pores, a negative pulse to deposit metal, followed by a delay to allow diffusion of ions. By optimising the pulse shape and duration, we have achieved uniform growth of metal into pores. Further, monitoring the deposition current helps to identify different phases in nano growth.

9.4 TRIBOLOGICAL CHARACTERISATION OF NPA-BASED NANOCOMPOSITE COATINGS

Nanoindentation of the porous alumina and the copper-filled nanocomposite showed that they have higher hardness than the aluminium substrate [20]. The hardness of the filled nanocomposite is about 40% higher than the corresponding unfilled porous alumina [12, 20]. Further, the nanocomposite had no circumferential cracks. The tribological properties of NPA and NPA-based composite coating were examined using an inhouse built reciprocating tribometer [51, 52]. The reciprocating friction experiments were performed on a ball-on-flat contact under dry condition, where a hard Zirconia ball of 6 mm diameter was used as a counter surface. The reason for choosing the Zirconia ball is due to its higher hardness (hardness 12–14 GPa) compared to the NPA. Hence, the ball will not wear out. In previous studies of tribological properties of NPA where the hardness of the counter surface was equal or less than the NPA coating resulted in worn of both surface and tribochemical reactions at the contacting interface, which influences the tribological properties [53, 54].

The schematic diagram of the modified tribometer is shown in Figure 9.7 [51]. The applied normal load was ranging from 0.98 N to 4.91 N, and the estimated Hertzian contact pressure varied from 700 MPa to 1200 MPa, which was well below the hardness of the sample. The normal loads were applied using dead weights. The friction force was measured using a high stiffness piezoelectric force sensor and real-time wear depth was measured using a highly sensitive non-contact fiberoptic displacement sensor at a sampling frequency of 1 kHz. All the experiments were performed at room temperature of 25°C and 50% relative humidity. The fiberoptic displacement sensor of 10 nm resolution was mounted on a Z-axis positioner with a micrometre stage fixed at the back side of the tribometer. The tip of displacement sensor was kept facing a reflecting gold mirror. The complete reciprocating system of the tribometer was pivoted, as shown in Figure 9.7. The normal load was applied by calibrated dead weights on the front side. As the samples wear out, the base plate moves downwards to the sample. Hence, the gap between the displacement sensor tip and the mirror decreases at the back side, which directly gives the wear depth of the sample. Thus, the initial contact was in elastic for all NPA coating samples. All experiments were done at a constant rate of 4.5 cycles/s and

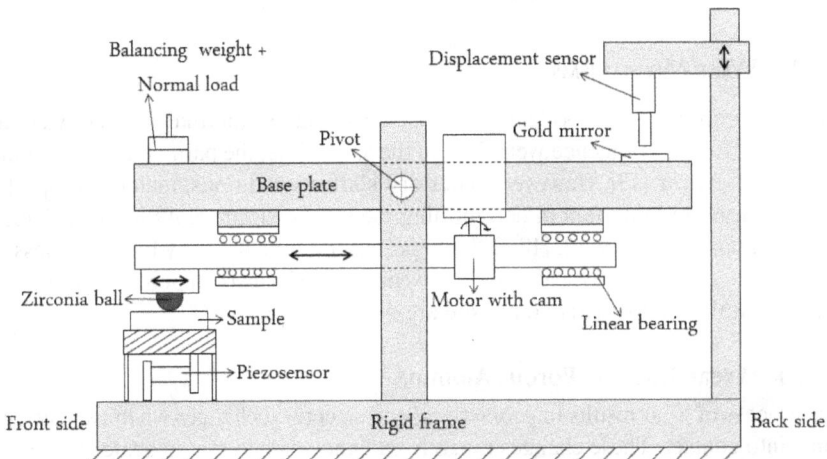

FIGURE 9.7   Schematic diagram of a ball-on-flat reciprocating tribometer [51, 52].

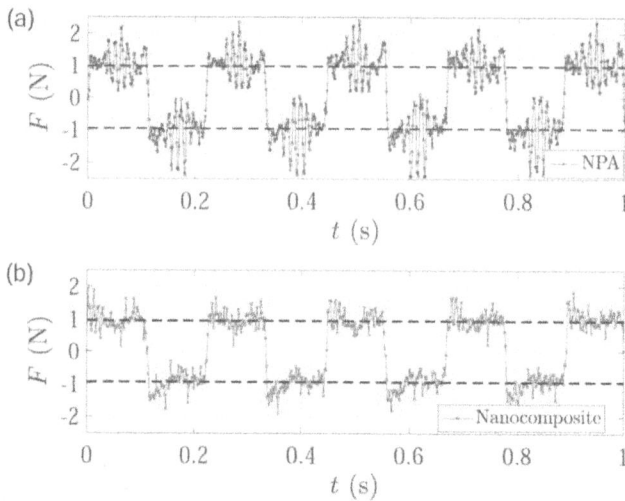

FIGURE 9.8 Typical friction force response as a function of time. (a) Friction force response of NPA and (b) for nanocomposite [13].

the stroke length of 1 mm. Before reciprocating tests, all samples and the counter surface were cleaned with an ultrasonic cleaner in deionised water for 5 min and then in ethanol for 10 min.

From the friction force response (Figure 9.8), the mean tangential force was obtained by taking a histogram of measured tangential force for every complete cycle [51]. The COF was calculated by dividing the mean tangential force with the applied normal load. The wear track of all samples and the counter surface (Zirconia ball) were examined using SEM. Figure 9.9 shows the variation of friction coefficient as the function of reciprocating cycles. The results show that nanocomposite has higher wear life compared to the NPA.

9.4.1 WEAR MECHANISMS

Material properties such as hardness, elastic modulus, thermal conductivity and thermal diffusivity influence wear. Yet, at the same time, the parameters that control wear are not clear [13]. However, under dry sliding conditions, material properties like hardness are important in determining the wear resistance. As we have demonstrated in our previous work [12, 20], the porous alumina has very high hardness. In this section, results from the dry sliding wear test performed on porous alumina and porous alumina/copper nanocomposite are discussed.

9.4.1.1 Wear Track on Porous Alumina

The process of wear results in geometrical and structural changes when two surfaces come into contact. These changes range from nano-scale to macro-scale [13]. At the nano-scale, asperities get reduced under sliding, whereas at the micro-scale, cracks

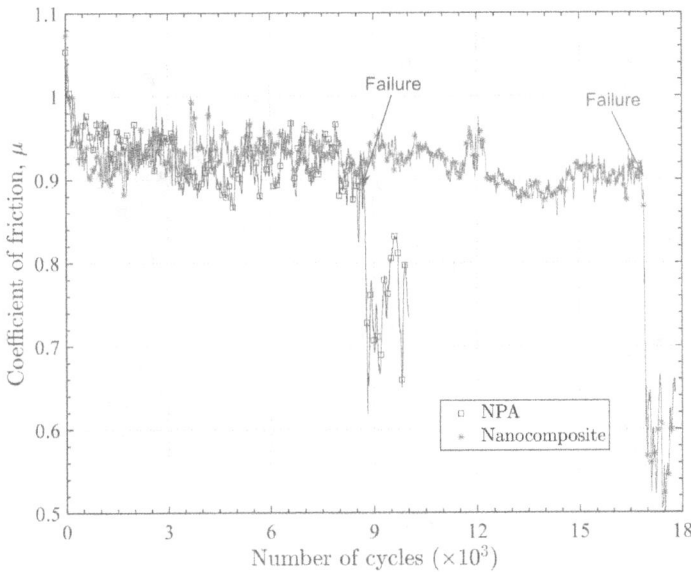

FIGURE 9.9 Coefficient of friction as a function of reciprocating cycles for NPA and nanocomposite [13].

get generated and debris is released. At the macro-scale, wear debris is agglomerated, and a surface layer is formed and deformed. However, the nanotextured structure of porous alumina can provide good wear resistance due to the high hardness of porous alumina. In this section, we study the wear characteristics of porous alumina.

Figure 9.10 shows the SEM image of wear tracks of porous alumina after a normal load of 300 g was applied under sliding against the stainless-steel flat substrate. Numerous cracks are observed on the surface (Figure 9.10a), which may have been caused due to the brittle nature of porous alumina. The porosity and brittle nature of porous Aluminium may give rise to cracks when subjected to tangential loads caused by sliding. The enlarged image near the crack reveals that pores still exist at the bottom (Figure 9.10b). This shows that the cracks are forming only on the top layer while the inner subsurface of the porous film still exists without delamination. The transfer layer can be a combination of porous alumina and stainless-steel substrate. The crack formation and propagation that happens only at the top layer might be due to the small penetration depth of the porous alumina against the steel substrate.

The enlarged partial portion of the wear track area of porous alumina in Figure 9.10c shows a compacted layer kind of structure with sparsely distributed wear debris. The wear debris found adhered inside the wear track area may have been formed due to a combination of abrasive and adhesive wear and may result in high values of friction coefficients. The sliding wear response in Figure 9.10d shows interesting features with few pores found to be filled with wear debris. The filling of the pores may have occurred due to the tiny abrasive particles formed just a few cycles after the commencement of sliding. The above discussion shows that when

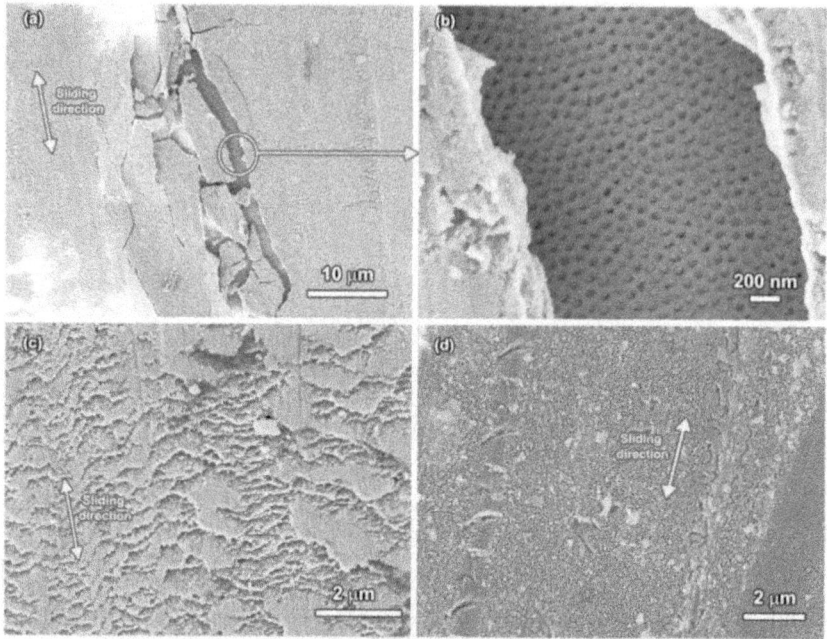

FIGURE 9.10 SEM images of wear track of (a) porous alumina and (b), (c) and (d) are the enlarged image of porous alumina at three different locations.

porous alumina is subjected to sliding under the applied normal load, even though crack formation and propagation occurs, the cracks are restricted only to the top layers. This shows that porous alumina is strong enough to withstand the contact pressure applied under sliding.

With a very hard porous alumina film on a softer aluminium substrate, it has been observed in the above section that at the given load, cracks formation occurs on the top layer. The coefficient of friction was also found to be high. A combination of brittle and ductile coating may prove to be more beneficial. For example, in the combination of ductile and brittle DLC/Carbide multilayered coating, the elastic layer allows the brittle layer to slide over each other in a manner of a multi-leaf book when bent [13]. The combination of ceramic/metal nanocomposites can improve wear characteristics as it provides one hard phase and one soft phase [1, 2]. Here we are studying the wear mechanism of porous alumina/copper nanocomposite coating, which is textured within the matrix. In the following section, we have studied the wear mechanism of this coating at both low and high coating thickness.

Figure 9.11 shows the SEM image of the wear track of porous alumina/copper nanocomposite of the low thickness (~1 μm) sliding against stainless steel after applying a normal load of 300 g. The wear mechanism observed is different from that of porous alumina. Inside the wear tracks, very little wear debris and fewer cracks are observed. In Figure 9.11a, a transfer layer is observed, which could be the combination of porous alumina/copper nanocomposite and stainless steel (Figure 9.11a).

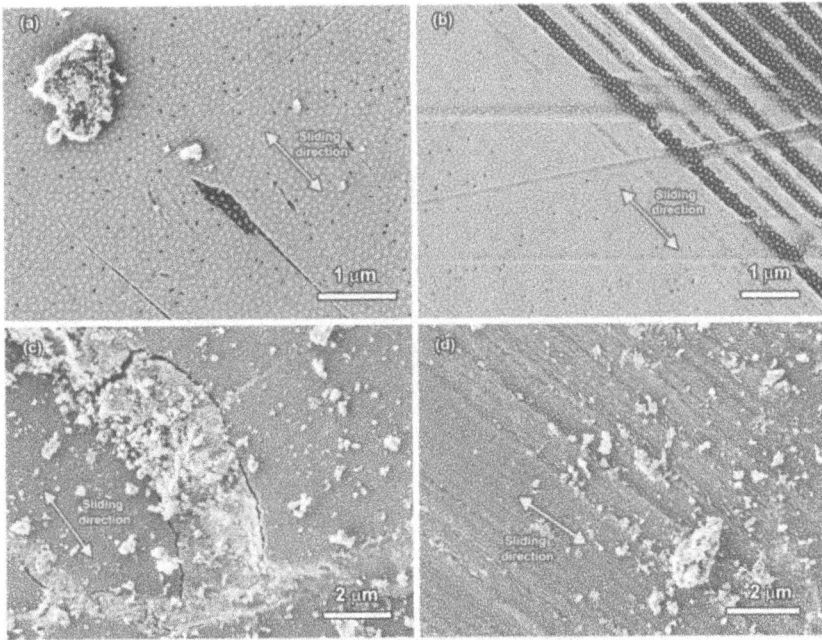

FIGURE 9.11 SEM images of wear track of (a) porous alumina/copper nanocomposite at a low thickness (~1 μm) and (b), (c) and (d) are the enlarged image of porous alumina/copper nanocomposite at three different locations.

At the top, a few scratches are observed, which might have been caused due to a hard abrasive particle entrapped in between the contact during sliding. However, these scratches confirm that a transfer layer exists. However, the surface beneath the top layer seems to be unaffected by the sliding.

A wavy surface morphology with peak and valley is observed in Figures 9.11b and 9.11d. This might be because under dry conditions, abrasive wear occurs during sliding. The hard abrasive wear particles formed could be a reason for the multitude of scratches observed on the transfer layer. In Figure 9.11c, some cracks are observed. The small wear particles that form during the initial cycles after the commencement of sliding would agglomerate to form larger particles in the subsequent cycles. These large sharp and hard particles rub against the steel substrate during sliding, which eventually results in the generation of the crack. Usually, geometry-related parameters like surface roughness, film thickness, hardness and wear debris influence the wear mechanism of the coating. In this section, we use high thickness (~3 μm) of porous alumina/copper nanocomposite coating for sliding. The wear track image of porous alumina/copper nanocomposite at high thickness sliding against steel is shown in Figure 9.12. From the figure, the surface is completely free from cracks. The wear debris formation is almost negligible. As sliding ensued, some Cu debris forms on the surface. This Cu debris acts as a lubricating medium between the coating and steel and prevents not only the wear of the coating but also the crack

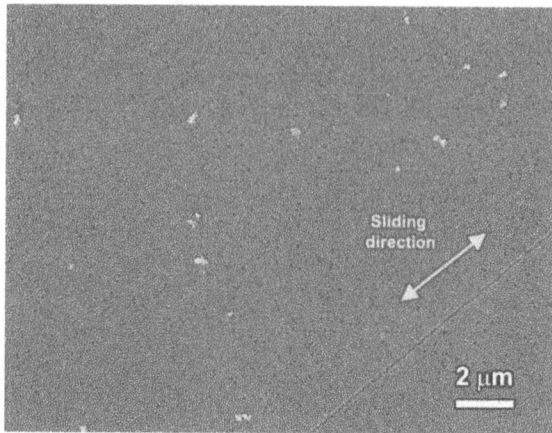

FIGURE 9.12 SEM images of wear track of porous alumina/copper nanocomposite at a high thickness (~3 μm).

formation. The white spots on the surface seen in the image may be the Cu debris particles. Hence, we find that porous alumina/copper-based nanocomposite coating proves to be extremely effective in improving the abrasive wear resistive coating under dry conditions.

9.5 CONCLUSION

In summary, a novel nanocomposite coating with nanoporous alumina as a matrix with aligned metal nanorod has been developed. This was achieved by optimally modifying the BL without sacrificing the interfacial strength. For filling metal into the porous structure, we have tried three different electrodeposition processes: DC, AC and pulse electrodeposition. We have found that the use of pulse electrodeposition is a highly efficient and well-suited method for metal filling into the pores. After developing the uniform nanocomposite coating, we have evaluated the tribological performance of the nanocomposite coating by measuring friction and wear in a reciprocating wear test. The coefficient of friction for NPA and Al samples were high, while it was 30% less for NPA/Cu. In NPA, crack formation occurs at the top layer, but the sub-layers remain unaffected. From the SEM images, wear and crack were found to be less in porous alumina/copper than in porous alumina and aluminium. A thin transfer layer of copper/steel gets coated on alumina. At a low thickness of porous alumina/copper nanocomposite coating, few cracks are found on the wear track area. At a high thickness of porous alumina/copper nanocomposite coating, no cracks are found. By optimising the nanoporous structure and tuning the electrodeposition process, we uniformly filled ordered pores with copper. Uniform coating has been achieved over an area. The coating is found to have good tribological properties of low friction and high wear resistance. Uniform coating has been achieved over an area of 10 mm × 10 mm. The coating is found to have higher hardness and higher wear resistance.

REFERENCES

1. Musil J 2000 Hard and superhard nanocomposite coatings *Surf. Coat. Technol.* **125** 322–330
2. Musil J, Zeman P, Hrubỳ H and Mayrhofer P 1999 ZrN/Cu nanocomposite film—a novel superhard material *Surf. Coat. Technol.* **120** 179–183
3. Alkire R C, Gogotsi Y, Simon P and Eftekhari A 2008 *Nanostructured materials in electrochemistry* (John Wiley & Sons)
4. Friedrich C, Berg G, Broszeit E and Berger C 1998 Fundamental economical aspects of functional coatings for tribological applications *Surf. Coat. Technol.* **98** 816–822
5. Merlo A M 2003 The contribution of surface engineering to the product performance in the automotive industry *Surf. Coat. Technol.* **174–175** 21–26
6. Ernst P and Barbezat G 2008 Thermal spray applications in powertrain contribute to the saving of energy and material resources *Surf. Coat. Technol.* **202** 4428–4431
7. Andrievski R A 1994 Nanocrystalline high melting point compound-based materials *J. Mater. Sci.* **29** 614–631
8. Cremer R, Reichert K, Neuschütz D, Erkens G and Leyendecker T 2003 Sputter deposition of crystalline alumina coatings *Surf. Coat. Technol.* **163–164** 157–163
9. Erdemir A 2005 Review of engineered tribological interfaces for improved boundary lubrication *Tribol. Int.* **38** 249–256
10. Voevodin A A and Zabinski J S 2005 Nanocomposite and nanostructured tribological materials for space applications *Compos. Sci. Technol.* **65** 741–748
11. Veprek S and Veprek-Heijman M J G 2008 Industrial applications of superhard nanocomposite coatings *Surf. Coat. Technol.* **202** 5063–5073
12. Yadav A 2015 *Nano Porous Alumina Based Composite Coating for Tribological Applications* PhD Thesis, IISc Bangalore.
13. M M, Yadav A and Bobji M S 2020 Wear characteristics of nanoporous alumina and copper filled nanocomposite coatings *Wear* **462–463** 203496
14. Stachowiak G and Batchelor A W 2013 *Engineering tribology* (Butterworth-Heinemann)
15. Erdemir A and Donnet C 2006 Tribology of diamond-like carbon films: recent progress and future prospects *J. Phys. Appl. Phys.* **39** R311–R327
16. Moll E and Bergmann E 1989 Hard coatings by plasma-assisted PVD technologies: Industrial practice *Surf. Coat. Technol.* **37** 483–509
17. Choy K L 2003 Chemical vapour deposition of coatings *Prog. Mater. Sci.* **48** 57–170
18. Pshyk A V, Coy L E, Nowaczyk G, Kempiński M, Peplińska B, Pogrebnjak A D, Beresnev V M and Jurga S 2016 High temperature behavior of functional TiAlBSiN nanocomposite coatings *Surf. Coat. Technol.* **305** 49–61
19. Safranek W H and Reed A H 1975 The properties of electrodeposited metals and alloys. A handbook *J. Electrochem. Soc.* **122** 270C
20. Durney L J 1984 *Graham's electroplating engineering handbook* (Springer Science & Business Media)
21. Yadav A, Muthukumar M and Bobji M S 2016 Porous alumina based ordered nanocomposite coating for wear resistance *Mater. Res. Express* **3** 085021
22. Parkhutik V and Shershulsky V 1992 Theoretical modelling of porous oxide growth on aluminium *J. Phys. Appl. Phys.* **25** 1258
23. Lei Y, Cai W and Wilde G 2007 Highly ordered nanostructures with tunable size, shape and properties: A new way to surface nano-patterning using ultra-thin alumina masks *Prog. Mater. Sci.* **52** 465–539
24. Hoar T P and Mott N F 1959 A mechanism for the formation of porous anodic oxide films on aluminium *J. Phys. Chem. Solids* **9** 97–99
25. Li F, Zhang L and Metzger R M 1998 On the growth of highly ordered pores in anodized aluminum oxide *Chem. Mater.* **10** 2470–2480

26. Siejka J and Ortega C 1977 An O18 study of field-assisted pore formation in compact anodic oxide films on aluminum *J. Electrochem. Soc.* **124** 883–891
27. Thompson G E and Wood G C 1981 Porous anodic film formation on aluminium *Nature* **290** 230–232
28. Su Z and Zhou W 2008 Formation mechanism of porous anodic aluminium and titanium oxides *Adv. Mater.* **20** 3663–3667
29. Thompson G 1997 Porous anodic alumina: Fabrication, characterization and applications *Thin Solid Films* **297** 192–201
30. Diggle J W, Downie T C and Goulding C 1969 Anodic oxide films on aluminum *Chem. Rev.* **69** 365–405
31. Masuda H and Fukuda K 1995 Ordered metal nanohole arrays made by a two-step *Science* **268** 1466–1468
32. Nielsch K, Müller F, Li A-P and Gösele U 2000 Uniform nickel deposition into ordered alumina pores by pulsed electrodeposition *Adv. Mater.* **12** 582–586
33. AlMawlawi D, Coombs N and Moskovits M 1991 Magnetic properties of Fe deposited into anodic aluminum oxide pores as a function of particle size *J. Appl. Phys.* **70** 4421–4425
34. Li F, Metzger R M and Doyle W 1997 Influence of particle size on the magnetic viscosity and activation volume of α-Fe nanowires in alumite films *Magn. IEEE Trans. On* **33** 3715–3717
35. Routkevitch D, Bigioni T, Moskovits M and Xu J M 1996 Electrochemical fabrication of CdS nanowire arrays in porous anodic aluminum oxide templates *J. Phys. Chem.* **100** 14037–14047
36. Routkevitch D, Chan J, Xu J and Moskovits M 1997 Porous anodic alumina templates for advanced nanofabrication *Proceedings of the International Symposium on Pits and Pores: Formation, Properties, and Significance for Advanced Luminescent Materials* vol 97 pp 350–357
37. Yadav A, Bobji M S and Bull S J 2019 Controlled growth of highly aligned Cu nanowires by pulse electrodeposition in nanoporous alumina *J. Nanosci. Nanotechnol.* **19** 4254–4259
38. Rabin O, Herz P R, Lin Y-M, Akinwande A I, Cronin S B and Dresselhaus M S 2003 Formation of thick porous anodic alumina films and nanowire arrays on silicon wafers and glass *Adv. Funct. Mater.* **13** 631–638
39. Tian M, Xu S, Wang J, Kumar N, Wertz E, Li Q, Campbell P M, Chan M H and Mallouk T E 2005 Penetrating the oxide barrier in situ and separating freestanding porous anodic alumina films in one step *Nano Lett.* **5** 697–703
40. Zhao X, Seo S-K, Lee U-J and Lee K-H 2007 Controlled electrochemical dissolution of anodic aluminum oxide for preparation of open-through pore structures *J. Electrochem. Soc.* **154** C553–C557
41. Martin C R 1996 Membrane-based synthesis of nanomaterials *Chem. Mater.* **8** 1739–1746
42. El-Sayed M A 2001 Some interesting properties of metals confined in time and nanometer space of different shapes *Acc. Chem. Res.* **34** 257–264
43. Choi J, Sauer G, Nielsch K, Wehrspohn R B and Gösele U 2003 Hexagonally arranged monodisperse silver nanowires with adjustable diameter and high aspect ratio *Chem. Mater.* **15** 776–779
44. Xu W, Chen H, Zheng M, Ding G and Shen W 2006 Optical transmission spectra of ordered porous alumina membranes with different thicknesses and porosities *Opt. Mater.* **28** 1160–1165
45. Cheng W, Steinhart M, Gösele U and Wehrspohn R B 2007 Tree-like alumina nanopores generated in a non-steady-state anodization *J. Mater. Chem.* **17** 3493

46. Shaban M, Hamdy H, Shahin F, Park J and Ryu S-W 2010 Uniform and reproducible barrier layer removal of porous anodic alumina membrane *J. Nanosci. Nanotechnol.* **10** 3380–3384

47. Trahey L, Becker C R and Stacy A M 2007 Electrodeposited bismuth telluride nanowire arrays with uniform growth fronts *Nano Lett.* **7** 2535–2539

48. Gerein N J and Haber J A 2005 Effect of ac electrodeposition conditions on the growth of high aspect ratio copper nanowires in porous aluminum oxide templates *J. Phys. Chem. B* **109** 17372–17385

49. Sun M, Zangari G and Metzger R M 2000 Cobalt island arrays with in-plane anisotropy electrodeposited in highly ordered alumite *Magn. IEEE Trans. On* **36** 3005–3008

50. Sauer G, Brehm G, Schneider S, Nielsch K, Wehrspohn R B, Choi J, Hofmeister H and Gösele U 2002 Highly ordered monocrystalline silver nanowire arrays *J. Appl. Phys.* **91** 3243–3247

51. M Muthukumar 2017 *Effect of Surface Texture on Friction under Dry Reciprocating Contacts* PhD Thesis, Indian Institute of Science, Bangalore.

52. M Muthukumar, M S Bobji 2018 Effect of micropillar surface texturing on friction under elastic dry reciprocating contact *Meccanica* **53** (9) 2221–2235

53. H S Kim, D H Kim, W Lee, S J Cho, J H Hahn and H S Ahn 2010 Tribological properties of nanoporous anodic aluminum oxide film, *Surf. Coat. Technol.* **205** (5) 1431–1437

54. N Tsyntsaru, B Kavas, J Sort, M Urgen and J P Celis 2014 Mechanical and frictional behaviour of nano-porous anodised aluminium *Mater. Chem. Phys.* **148** (3) 887–895

Index

3D printer, 115, 116
5G, 4, 21

A

Additive manufacturing, 27, 52–56, 100, 110, 156, 157
Advanced manufacturing, i, vii, viii, 28, 29, 37, 51, 52
AI, 2
Antibiofilm, 159
Anti-pathogenic, 27

B

Big Data, vii, 1–3, 5–7, 9, 11, 13, 15, 17, 19–21, 23, 25
Build time, 114, 133–135

C

Carbon, 38, 54, 159, 161, 171–174
Carbon nanotubes, 9, 159–161, 163–166, 168–170
CNT-based composites, 161, 164, 166, 168, 170
Cold spray, vii, 28–34, 36–52, 63, 76
Commercial Modular Aero-Propulsion System Simulation, 2, 14–16, 19
Convolutional Neural Network, 7, 9, 14, 15
COPRAS, viii, 57, 58, 61, 63–66, 77, 79
CPS, 3, 4, 5, 8
CSAM, 46–50, 52, 54

E

EDM, viii, ix, 61–63, 67, 68, 70, 72, 79, 80, 84, 91, 99, 100, 101–111
EHSV, viii, 83–86, 98, 99
Extrusion temperature, 113, 114, 118, 121, 122, 125, 126, 128, 130, 134

F

FDM process, 114, 116, 117, 137, 140, 141
FDM technology, 113
Full factorial design, 103

G

Gas temperature, 27, 34, 52, 54
Green applications, 27
Grey relational generation, 91, 92

Grey relational grade, 84, 97
GRPA, xi, 116, 118, 119, 121–124, 126–128, 130–133, 137

H

Higher the better, 90, 91

I

IIoT, 1–5, 9–14, 19, 20
Industry 4.0, vii, 2, 3, 6–9, 11, 13, 14, 20–25
Internet of Things, 3

K

KRPA, 116, 118, 119, 121–126, 129–133, 137

L

LPCS, 30, 31, 33, 44, 54
LPWA, 4, 21

M

Machine-to-Machine Communication, 3, 4, 13
Machining Techniques, 57
Manufacturing, i, xi, 1–26, 27, 28, 30, 32, 34, 36, 38, 40, 42, 44, 46, 48, 50, 52, 53, 54, 55, 56, 58, 60, 62, 64, 66, 68, 70, 72, 74, 76, 78, 79–82, 84, 86, 88, 90, 92, 94, 96, 98, 100, 102, 104, 106, 108, 110, 113, 114, 116, 118, 120, 122, 124, 126, 128, 130, 132, 134, 136, 138, 140, 142, 144, 146, 148, 150, 152, 154, 156, 157, 159, 160, 161, 164, 166, 168, 170, 171, 172, 173, 174, 175, 176, 178, 180, 182, 184, 186, 188, 190, 192
Manufacturing method, i, 2, 4, 6, 8, 10, 12, 14, 16, 18, 20, 22, 24, 26, 27, 28, 30, 32, 34, 36, 38, 40, 42, 44, 46, 48, 50, 52, 54, 56, 58, 60, 62, 64, 66, 68, 70, 72, 74, 76, 78, 80, 82, 84, 86, 88, 90, 92, 94, 96, 98, 100, 102, 104, 106, 108, 110, 114, 116, 118, 120, 122, 124, 126, 128, 130, 132, 134, 136, 138, 140, 142, 144, 146, 148, 150, 152, 154, 156, 160, 164, 166, 168, 170, 172, 174, 176, 178, 180, 182, 184, 186, 188, 190, 192
Manufacturing process, i, 2, 3, 8, 9, 12, 13, 14, 19, 45, 50, 51, 137

Material removal rate, 61–63, 65–68, 70, 72, 74, 77, 83, 85, 88–99, 101

MCDM, viii, 57, 58, 59, 60, 63, 67, 69, 72, 73, 74, 75, 77, 78, 79, 80, 81, 140

Mechanical manufacturing, 2

Morphology, 27, 53, 56

Multi-Attributive Border Approximation Area Comparison, viii, 57, 58, 76, 77, 81, 82

Multi-Response Optimization, 84

Multi-wall carbon nanotubes, 159, 161–170, 172

N

Nanocomposite, vi, 175, 177, 179, 181, 183, 185, 187, 189, 191, 193

Nanoporous alumina, 9, 175, 176, 177, 178, 180, 184, 185, 186, 187, 190

Nanotubes, 6, 159, 161, 171, 172, 173

Nozzle design, 27

Nozzle scanning, 27, 34

O

Optimization, i, vii, 3, 6, 12, 23, 24, 25, 61, 63, 66, 68, 77, 78, 79, 80, 81, 82, 85, 91, 96, 99, 100, 103, 104, 107, 109, 110, 115, 140, 141, 146, 157

P

Predictive analytics, 7, 8

Print speed, 113, 114, 118, 121, 123, 125, 128, 131, 134

Processing parameters, 84, 118

Product lifecycle management, 10, 11

R

Repair applications, 27

RFID, vii, 3, 4, 5, 10, 20

S

SENSORS, 5

Single-wall carbon nanotubes, 159, 168, 169

Smart Processes, i

Spool bore, 83

Spray angle, 27, 35, 54

Stainless steel, 53, 83, 86, 143, 148

Standoff distance, 27, 35, 54, 55

Statistical process monitoring, 9

Supply chain, 3, 20

T

Taguchi-based GRA, 83, 85, 87, 89, 91, 93, 95, 97, 99

Tensile, 113, 116, 120, 128, 129, 137, 141

Thermal spray technology, 27, 52–56

Topological Data Analysis, 9

Total cost, 113, 114, 120, 133, 134, 136

Tribological, 6, 32, 52, 53, 54, 175, 191, 193

V

VIKOR, viii, 57, 58, 60, 61, 62, 63, 77, 78, 79

Virtual reality, 12, 13

Volume, Velocity, and Variety, 6

W

Wearing behaviour, 101, 103, 105, 107, 109, 111

WEDM, viii, 67, 68, 70, 74, 76, 79, 81, 83, 84, 85, 87, 94, 96, 98, 99, 100

White layer thickness, 63

For Product Safety Concerns and Information please contact our EU
representative GPSR@taylorandfrancis.com
Taylor & Francis Verlag GmbH, Kaufingerstraße 24, 80331 München, Germany